Ivan Gutman · Oskar E. Polansky

Mathematical Concepts in Organic Chemistry

With 28 Figures

Springer-Verlag
Berlin Heidelberg NewYork
London Paris Tokyo

Professor **Ivan Gutman**
Faculty of Science
P.O. Box 60
YU-34000 Kragujevac

Professor **Oskar E. Polansky**
Max-Planck-Institut für Strahlenchemie
Stiftstr. 34–36
D-4330 Mülheim a. d. Ruhr

ISBN-13: 978-3-642-70984-5   e-ISBN-13: 978-3-642-70982-1
DOI: 10.1007/978-3-642-70982-1

Library of Congress Cataloging-in-Publication Data
Gutman, Ivan, 1947 –
Mathematical concepts in organic chemistry.
Bibliography: p.
Includes index.
1. Chemistry, Organic — Mathematics. I. Polansky, Oskar E., 1919 –
QD255.5.M35G87 1986 547′.001′51 86-13103
ISBN 0-387-16235-6 (U.S.)

2152/3020-543210

Preface

The present book is an attempt to outline some, certainly not all, mathematical aspects of modern organic chemistry. We have focused our attention on topological, graph-theoretical and group-theoretical features of organic chemistry, Parts A, B and C.

The book is directed to all those chemists who use, or who intend to use mathematics in their work, and especially to graduate students. The level of our exposition is adjusted to the mathematical background of graduate students of chemistry and only some knowledge of elementary algebra and calculus is required from the readers of the book. Some less well-known, but still elementary mathematical facts are collected in Appendices 1–4. This, however, does not mean that the mathematical rigor and numerous tedious, but necessary technical details have been avoided. The authors' intention was to show the reader not only how the results of mathematical chemistry look, but also how they can be obtained.

In accordance with this, Part D of the book contains a few selected advanced topics which should give the reader the flavour of the contemporary research in mathematical organic chemistry.

One of the authors (I.G.) was an Alexander von Humboldt fellow in 1985 when the main part of the book was written. He gratefully acknowledges the financial support of the Alexander von Humboldt Foundation which enabled his stay at the Max-Planck-Institut für Strahlenchemie in Mülheim and the writing of this book.

The authors are indebted to Dr. HEINZ BARENTZEN and Professor MAXIMILIAN ZANDER for valuable discussions.

The manuscript of the book was able to be completed thanks to the enthusiastic assistence of Mrs. INGEBORG HEUER. The authors also thank Mrs. RENATE SPECKBRUCK and Mrs. EVELINE CURRELL for their help.

Mülheim a. d. Ruhr, April 1986   **Ivan Gutman   Oskar E. Polansky**

Contents

Chapter 9 Automorphism Groups 108

Chapter 10 Some Interrelations between Symmetry and Automorphism Groups 117

Part D Special Topics 121

Chapter 11 Topological Indices 123

Chapter 12 Thermodynamic Stability of Conjugated Molecules. 135

Introduction

Mathematical chemistry nowadays presents a variety of approaches to understanding the mathematical structures which lie behind existing chemical concepts, to establishing and investigating novel mathematical models of chemical phenomena, and applying mathematical ideas and techniques in chemistry. Throughout the entire history of chemistry certain scientists, usually not numerous, were inclined to contemplate connections between mathematics and chemistry and the possibility of using mathematics for deducing known and predicting new chemical facts. Extensive use of mathematical methods is traditional in various branches of physical chemistry, especially in thermodynamics (partial derivatives, proper and improper differentials, path integrals etc.) and chemical kinetics (coupled non-linear differential equations). A real need for mathematics in chemistry appeared, however, only after the discovery, made by physicists in the first three decades of our century, that the basic properties of atoms and molecules can be explained and predicted by means of quantum theory. The awareness that chemistry cannot be understood without a knowledge of quantum physics, including its sophisticated mathematical apparatus, was the actual driving force which led to the introduction of mathematics and mathematical thinking into (or at least not very far from) chemical laboratories.

The advent of computers gave another impetus to the mathematization of chemistry. Computers not only enable calculations and data processing which were inconceivable in the recent past, but they are also stupid enough to require absolutely unequivocal, precise and complete instructions, thus forcing the programmer to think in a mathematician's manner.

The constitution of mathematical chemistry as a novel specialized and interdisciplinary branch of Science is just a necessary consequence of the mathematization which is nowadays occurring in chemistry[1]. This, of course, does not mean that certain parts of mathematical chemistry, especially those related to mathematical analysis (derivatives, integrals, differential equations) have not been well established for many years. For a long time mathematics was imported into chemistry from physics and, as a consequence of this, those parts of mathematical chemistry which are of little or no importance for physicists remained underdeveloped. This especially applies to various topological and combinatorial aspects of (organic) chemistry.

A rapid proliferation of genuine mathematical investigations in organic chemistry began somewhere in the early seventies, when the significance of graph theory became

---

[1] A similar process is currently under way in biology whereas mathematical physics has a tradition of some two centuries.

fully recognized. Progress in mathematical organic chemistry is clearly illustrated by the great number of newly obtained results. It can be estimated that after 1979 more than 600 relevant papers have appeared in the chemical literature. The first scientific meeting on mathematical chemistry (chemical graph theory) was held in 1975. Some ten such conferences have since been organized. The journal Mathematical Chemistry (Match) has been published since 1975. For a list of existing books and reviews on mathematical chemistry and related topics see the bibliography at the end of this book.

The present book is an attempt to give a cross-section through the contemporary mathematical research on organic molecules. It was not the authors' aim to offer a complete or even a nearly complete review of the plethora of approaches which have appeared in the literature in the last 10–15 years. Rather we shall focus our attention on a few selected topics and elaborate them in some detail. This selection, however, has been made in such a manner as to provide a representative image of the present status of the mathematical chemistry of organic molecules.

The book is mainly concerned with theorems (i.e. non-trivial statements which are mathematically correct and which have been proved in the mathematical sense of this word). Theorems represent the heart of any scientific theory which can be characterized as mathematical. The authors' intention was not only to outline the existing results, but also to demonstrate how they have been obtained. Complete proofs are presented only for a limited number of theorems. Otherwise only the idea of the proof is indicated or a pertinent reference is quoted. In general, however, the reader of this book will get a sufficient insight into the proof techniques of modern mathematical chemistry and may become able eventually to deduce his own results.

In addition to this formal approach, we illustrate our statements by concrete chemical examples and/or applications.

The book is divided into 13 chapters and 6 appendices. Chapters 1 and 2 are concerned with topological, Chapters 3 to 6 with graph-theoretical and Chapters 7 to 10 with group-theoretical issues in organic chemistry. Chapters 11 to 13 elaborate in more detail three selected topics, namely certain topological indices, the thermo-dynamic stability of conjugated molecules and the so-called topological effect on molecular orbitals. The Appendices 1 to 4 will remind the readers of the theory of matrices, determinants, eigenvalues and polynomials, respectively. Appendix 5 is closely related to Chapter 8 and contains the character tables of many chemically significant symmetry groups. The readers, especially those who are not starting to read the book from its beginning should not hesitate in consulting Appendix 6, where a list of symbols used is given together with a brief explanation of their meaning.

Part A

Chemistry and Topology

Chapter 1

Topological Aspects in Chemistry

> In constitutional formulas
> the atoms are represented
> by letters and the bonds
> by lines. They describe the
> topology of the molecule.
> V. PRELOG, Nobel Lecture, December 12, 1975 [165]

1.1 Topology in Chemistry

Since the time of VAN'T HOFF, chemists are used to thinking about molecules as geometric objects in which atoms have a certain spatial arrangement. The geometric parameters of molecules (interatomic distances, bond angles, dihedral angles, etc.) can be measured with a rather high degree of accuracy and are indeed known in a considerable number of cases.

It is, however, perfectly clear that the positions which the atoms occupy in the molecule are not fixed and various kinds of intramolecular motions (internal rotations, vibrations, conformational changes, etc.) are known to occur. Even if one disregards these atomic motions, the geometry of a molecule is to some extent influenced by its environment (e.g. by pressure in the case of crystals, by the choice of the solvent in the case of solutions, etc.). The difficulties in bringing the concept of molecular geometry in harmony with the basic principles of quantum theory are also worth mentioning (see [181] and the references cited therein).

On the other hand it is beyond dispute that the molecule maintains its identity irrespective of all these geometric changes. Hence there must be something in the molecule which remains invariant when (slight) modifications in its geometry occur.

Having all this in mind, it is not surprising that the knowledge of exact geometric parameters of molecules is usually of little value when problems which occur in chemical practice have to be solved. Chemists are very often satisfied with much less detailed information about the structure of molecules. This is called "constitution" or "connectedness" in some, while "form" or "shape" in other cases. Another expression, becoming more and more popular, is "topology" [22, 44], although the precise meaning of this word is somewhat obscure.

Another drawback of the geometric approach is that the geometric parameters provide a local characterization of the molecule (i.e. they establish relations between two, three or four atoms). On the other hand, certain spatial relations in molecules

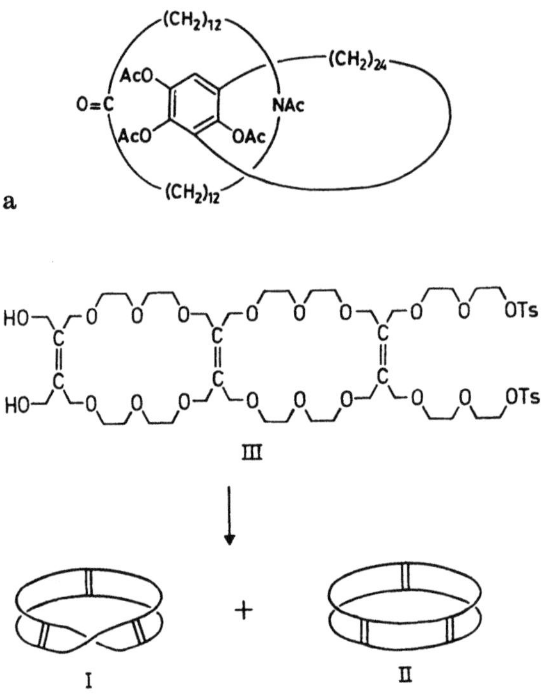

Fig. 1.1. **a** Example of a catenane [60]; **b** a MÖBIUS strip compound (I) and its normal ("cylindrical") isomer (II) obtained by cyclization of tris(tetrahydroxymethylene)diol ditosylate (III) [178]

are consequences of the molecular architecture as a whole. The most peculiar of such cases are the catenanes [60] and MÖBIUS compounds [178].

What is exceptional in these molecules are not their bond lengths and angles, but rather the embedding of the molecule as a whole in three-dimensional space [22].

The fact that certain properties of geometric objects remain invariant under continuous deformation[1] of their points has long been recognized. The theory of such phenomena was named "topology" and eventually became one of the most distinguished disciplines of modern mathematics [19]. Instead of conventional geometric objects (which are sets of points in the two- or three-dimensional space, whose distances are determined by the Euclidean metric), modern topology investigates sets with much more general properties.

A good part of the geometric problems considered in the early stages of topology have been overtaken by another mathematical discipline, namely graph theory. Graph theory seems to provide the real operative basis for the topological considerations in chemistry. The readers will become acquainted with chemical graphs in Chap. 3 and meet them thereafter very frequently.

Returning to geometric objects, we can say that two of them are topologically equivalent if one object can be continuously transformed so as to coincide with the other. For instance, the lines L_1 and L_2 are topologically equivalent whereas the line L_3 (possessing an articulation point A) is a topologically distinct entity.

---

[1] In topology instead of "continuous deformation" one speaks about "homeomorphic mapping", a concept which is then rigorously defined [19]. We use here a much more intuitive approach.

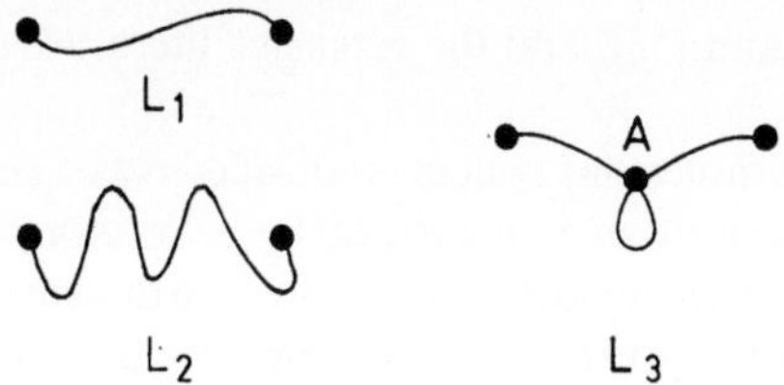

Similarly, there are two topologically different ways in which a pair of cycles can be embedded in three-dimensional space.

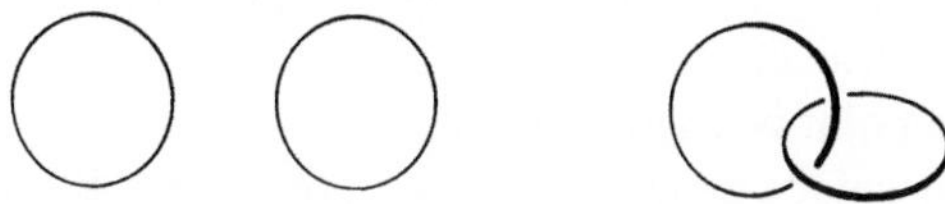

It is clear that any deformation which would transform the catenated cycles into disjoint ones would require the breaking of one of them. Such a deformation would not be continuous. It is less obvious (but can be demonstrated) that catenated and disjoint cycles can be continuously transformed into each other in four-dimensional space. This shows that the answer to the question whether two geometric objects are topologically equivalent depends not only on the objects themselves, but also on the properties of the space in which they are embedded. The reader should convince himself about this by considering the possible relations between two cycles embedded in two-dimensional space (i.e. in the plane).

Changes in the molecular geometry, caused by intramolecular motions or by any kind of external influence, but in which no chemical bond has been broken or formed, can be regarded as continuous deformations. Those characteristics of molecules which remain invariant under such deformations are therefore called *topological properties*.

The relations which exist between the topological and the real chemical and physical properties of the molecules are the subject of numerous recent mathematical investigations in organic chemistry. Some aspects of these studies are outlined in the present book.

At this point it should be clearly emphasized that the approach which we pursue in the present and the subsequent chapter is not the only way to introduce topological concepts into chemistry. First of all, throughout the previous discussion we have tacitly understood as granted that in a molecule there are *atoms*, that these atoms have a more or less fixed *position* in the molecule and that between some (but not all) atoms *chemical bonding* occurs. Hence our notion of molecular topology is based on classical chemical ideas and will be therefore "intuitively clear" to the readers of this book.

There exist, however, other, more sophisticated and mathematically much more rigorous approaches to molecular topology. These are obtained by using topological concepts within the quantum mechanical description of molecular systems. Two such theories, associated with the names of PAUL MEZEY and RICHARD BADER are especially interesting. Even a brief exposition of each of these theories would require a long chapter and a detailed mathematical introduction, which we cannot afford in the present book. The reader who wants to learn more about the work of MEZEY and BADER should consult a pertinent review (e.g. [4] and [53]) and the extensive literature cited therein.

In both MEZEY's and BADER's theories a molecular system is considered as a collection of electrons and atomic nuclei whose motion is described by a SCHRÖDINGER equation. If there are N nuclei in the system, then an abstract $n = 3N - 6$ dimensional nuclear configuration space $\mathscr{R}^n$ is introduced. MEZEY examines the potential energy hypersurface $E(R)$, $R \in \mathscr{R}^n$. The critical points of the energy hypersurface determine the number of possible reaction mechanisms of the given molecular system. Chemical structures are represented by the so-called *catchment regions*, which are open subsets of $\mathscr{R}^n$ and which include a critical point of $E(R)$. A stable molecule is represented by a catchment region of index $\lambda = 0$, i.e. the corresponding critical point of $E(R)$ is a minimum. A transition structure ("transition state") is represented by a catchment region of index $\lambda = 1$, whose critical point is a saddle point.

BADER's theory considers the molecular charge distribution $\varrho(r) = \varrho(r, R)$, where r is a three-dimensional position coordinate and $R \in \mathscr{R}^n$, as before. The basic property of $\varrho(r)$ is that it exhibits local maxima only at the positions of nuclei. The nuclei thus behave as point attractors in the associated gradient vector field $\nabla\varrho(r)$. According to BADER, an atom is defined as the union of an attractor and its basin. The border between two such atoms is again determined by certain critical points of $\varrho(r, R)$. The *atomic interaction line* is a path through space linking neighbouring nuclei, along which the charge density is a maximum with respect to any neighbouring path. A *molecular graph* is defined as the union of these interaction lines and such a graph may be assigned to each point R of the nuclear configuration space $\mathscr{R}^n$.

In topology, two vector fields $\mathscr{V}_1$ and $\mathscr{V}_2$ over $\mathscr{R}^3$ are said to be equivalent if there is a continuous one-to-one mapping of $\mathscr{R}^3$ into $\mathscr{R}^3$, which maps the trajectories of $\mathscr{V}_1$ into the trajectories of $\mathscr{V}_2$. BADER defines two nuclear configurations R_1 and R_2 to be equivalent if the vector fields $\nabla\varrho(r, R_1)$ and $\nabla\varrho(r, R_2)$ are topologically equivalent. Furthermore, two molecular graphs are defined to be equivalent if and only if they correspond to equivalent nuclear configurations. An equivalence class of molecular graphs is then called the *molecular structure*.

1.2 Abstraction in Science and How Far One Can Go

When a phenomenon occuring in Nature is studied from a scientific point of view then a convenient and rather fruitful procedure is to abandon the consideration of some of its details and focus the attention only on those features of the phenomenon which are expected to be significant. When such an abstraction is made in a reasonable way, then the loss of the knowledge of the details is abundantly compensated by the gain of a better insight into the phenomenon. The concept of ideal gas, absolutely black body, nucleon and the rationalization of the periodic system of elements by means of the Aufbau process are examples which come readily to mind.

Depending on the level of the abstraction, we may pose different sorts of questions as illustrated in Fig. 1.2 (see p. 10). Depending on the information we are interested in, we may choose an appropriate level of abstraction.

The following scheme suggests a possible classification of the ways in which the notion of a chemical substance and its structure can be viewed.

	Disregarded	Gained
Substance in free nature		
Substance in laboratory	Uncontrollable influences of the environment	Reproducibility
Chemically pure substance	Effect of impurities	Perfect reproducibility, quantitative measurements
Atomistic theory	Intermolecular interactions	Simple picture of chemical processes
Quantum theory	Non-quantum-mechanical (e.g. relativistic) effects	Explanation of basic chemical notions
Approximate quantum theoretical models	Several requirements of quantum mechanics	MO schemes VB schemes
Molecular geometry	Quantum mechanics	Applicability of Euclidean geometry to molecules
Structural formula	Details of chemical binding	Pictorial representation of molecular structure
Molecular topology	Differences between atoms	?

Molecular topology lies at the very bottom of our table, being thus the most abstract model for the structure of chemical substances. It has obviously lost a great deal of its relation to real chemistry. The question which naturally arises is whether by putting forward this model we did not go too far in the process of abstraction. In other words we are faced with the danger that molecular topology will not be able to reflect any meaningful property of chemical compounds.

The question mark in the right lower corner of the above table is related to the provocative question:

Is there any chemistry left in molecular topology?

The last three levels in our classification scheme can also be characterized in the following way:

	Disregarded	
Molecular geometry	Information about energy	Physics
Structural formula	Information about energy and space	Physics
Molecular topology	Information about energy, space and matter	No physics

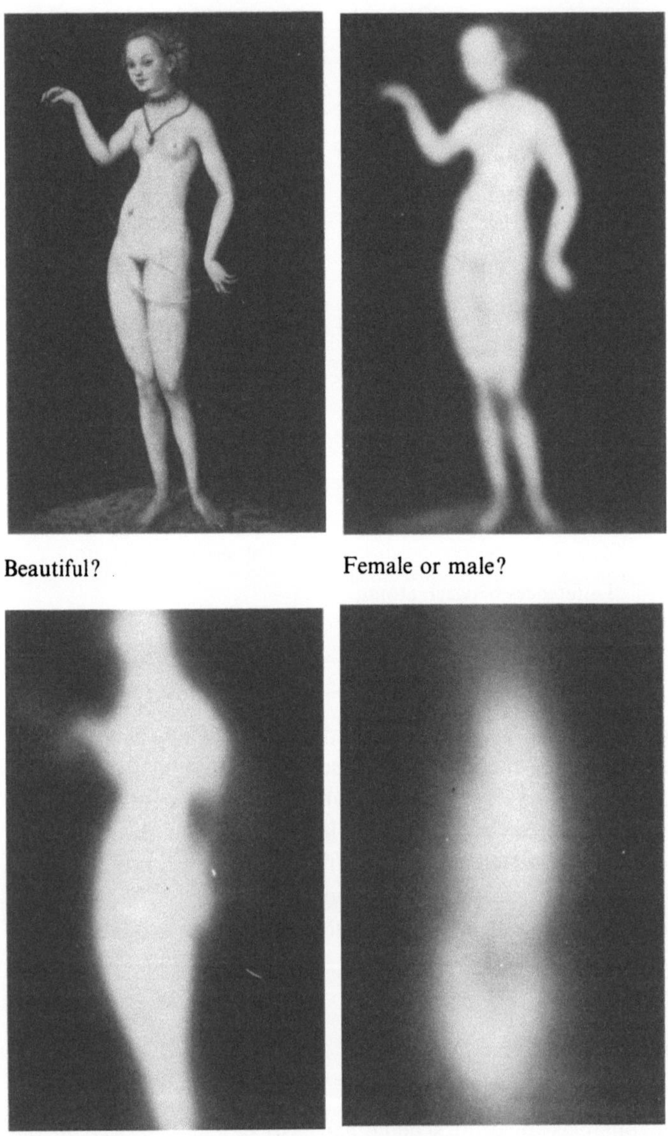

Beautiful?   Female or male?

Human or hare?   How many ?

Fig. 1.2. The sort of questions one may ask depends on the level of abstraction

Having in mind that natural sciences are concerned with the behaviour of matter in space and time, we are ready to conclude that on the level of molecular topology the connection with physical reality has been completely lost. This is definitely not so. As a matter of fact,

topology provides the frame for physics

and, in particular

molecular topology provides the frame for molecular physics and chemistry.

This frame has to be filled with a physical content, which shows it to be not fully independent of the frame itself.

Considerations based exclusively on molecular topology will give no results useful for chemistry[1]. It is a fortunate fact, however, that in a number of chemically quite important cases, molecular topology can be recoupled with certain physical models, thus enabling the estimation of both spatial and energetic relations. We shall elaborate this point in more detail later on and demonstrate the validity of our claim on a number of concrete examples.

---

[1] Topology suffices for enumeration and classification purposes, as a basis for a chemical nomenclature and similar applications.

Chapter 2

Molecular Topology

2.1 What is Molecular Topology?

A precise, but formal and not easily understandable definition of molecular topology will be given in Sect. 3 of this chapter. For most of the applications, however, the present simple description will fully suffice. Suppose a molecule M is composed of n atoms $A_1, A_2, \dots, A_n$. Suppose also that for any two atoms A_i and A_j we can decide which of the following two statements is correct:

(a) In the molecule M there is a chemical bond between the atoms A_i and A_j.

(b) In the molecule M the atoms A_i and A_j are not chemically bonded.

Then the totality of information about the mutual connectedness of all pairs of atoms in a molecule (*and only this information*) determines the topology of the respective molecule.

There are several ways in which the information on mutual connectedness of atoms in a molecule can be recorded. A diagramatical representation, which we shall describe now, can be easily grasped by organic chemists, because of its close resemblance to the structural formulas.

Represent each atom by a small circle and draw a line (not necessarily a straight one) between any two circles which correspond to chemically bonded atoms. The diagram obtained in this manner is a graph. Since it represents the molecular topology, it will be called a molecular graph.

Consider as an example ethanole and represent its topology as described above. For the sake of the reader's convenience we have labeled the atoms and the corresponding circles by 1, 2, ... , 9.

A chemist would obviously be more satisfied if the circles and lines were slightly rearranged:

There is no reason not to meet this aesthetic requirement, and in the following the diagramatic representation of the topology of an organic molecule will always be drawn so as to match the corresponding structural formula. It must, however, not be forgotten that this is just a matter of convenience. In particular, both graphs given above are equivalent and represent the very same molecular topology.

The concept of molecular graphs is explained in full detail in Chap. 3.

The idea of molecular topology and the degree of abstraction involved in it will become even more transparent after the definition of *isotopological molecules*. Hence, two molecules are said to be isotopological if their topologies coincide. On Fig. 2.1

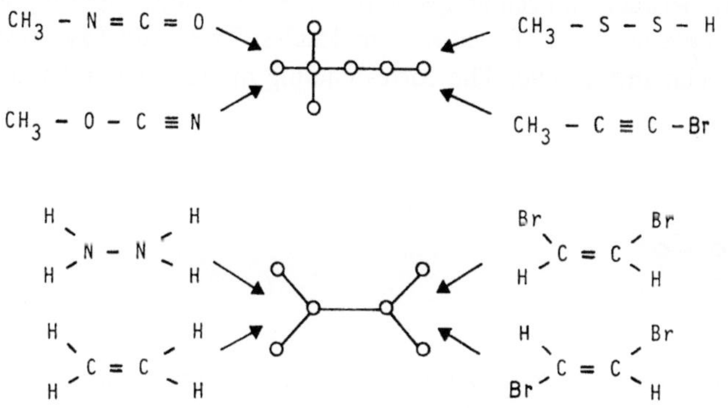

Fig. 2.1. Examples of isotopological molecules. These are indistinguishable from the point of view of molecular topology

are presented two quartets of isotopological molecules. These examples reveal an apparent difficulty in the topological approach, namely that one molecular graph represents a whole set of isotopological, but chemically distinct and often quite diverse compounds. Consequently if T is a certain topological property (which can be expressed by means of a number), then all members of an isotopological class have the same numerical value for T. Now, if P is a measurable physical quantity which is expected to be related with the molecular topology via a function

$$P = f(T, a, b, c, ...) \tag{1}$$

where $a, b, c, ...$ are parameters of the theory, then the function f must be the same for all classes of compounds and only the parameters $a, b, c, ...$ may vary from class to class.

At the present moment we know of a very limited number of molecular properties which would meet the above rigorous requirements. One of these seems to be a regularity (called TEMO), concerning the electron energies of certain topologically related isomers. More details on TEMO can be found in Chap. 13.

On the other hand, there is a lot of evidence that physical properties of particular classes of (organic) molecules can be reproduced (usually in a semiquantitative manner) from the molecular topology. Examples for this are given in Chaps. 11, 12 and elsewhere.

2.2 Geometry, Symmetry, Topology

The claim that molecular topology provides a frame for molecular physics and chemistry can be understood in the following manner. There are certain geometries which are compatible with a given molecular topology; all other molecular geometries are forbidden for topological reasons.

Symmetry is another important characteristic of a molecule. It depends both on molecular geometry and the type of atoms which form the molecule. This latter is denoted as the *materialization* of a certain molecular geometry or molecular topology.

The delicate interplay between molecular geometry, symmetry and topology is best illustrated by the simple example of triatomic molecules. These can have two different topologies — cyclic and acyclic. The corresponding molecular graphs are G_1 and G_2.

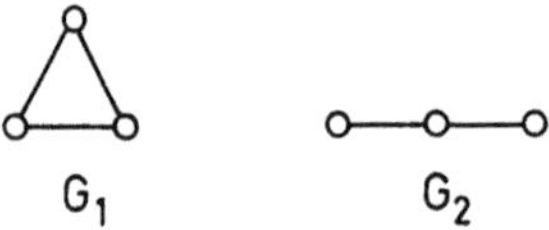

Table 2.1 presents the possible molecular geometries, their materializations and the corresponding symmetry groups. (For details on symmetry groups see Part C of the present book.)

Table 2.1. Topologies, geometries and symmetries in triatomic molecules and their mutual relationship

Topology	Geometry	Materialization			
		AAA	ABA	AAB	ABC
Cyclic (S_3)	Planar	D_{3h}		C_{2v}	C_{1h}
Acyclic	Planar	C_{2v}	C_{2v}	C_{1h}	C_{1h}
$(S_2 \oplus S_1)$	Collinear	$D_{\infty h}$	$D_{\infty h}$	$C_{\infty h}$	$C_{\infty v}$

For a more elaborated and mathematically rigorous treatment of these problems the reader should consult the work of ERNST RUCH on chirality functions [52, 167, 168].

2.3 Definition of Molecular Topology

The first attempt to give a precise definition of molecular topology using the concept of the molecular graph as a starting point seems to be the work of MERRIFIELD and

SIMMONS [145]. We shall elaborate here a somewhat different approach to the same problem. We assume that the reader of this section is already familiar with the concept of distance in a graph (see Sect. 4.1.4.).

Let us start with a few definitions [19].

Consider a set $\mathscr{X}$. Another set $\mathscr{X}_1$ is called a subset of $\mathscr{X}$ if every element of $\mathscr{X}_1$ is also an element of $\mathscr{X}$. This fact is denoted by $\mathscr{X}_1 \subseteq \mathscr{X}$. Let $\mathscr{X}_1, \mathscr{X}_2, \mathscr{X}_3, \ldots$ be subsets of the set $\mathscr{X}$.

An *open set topology* or an O-topology in the set $\mathscr{X}$ is a collection $\mathsf{T}_O = \{\mathscr{X}_1, \mathscr{X}_2, \mathscr{X}_3, \ldots\}$ of subsets of $\mathscr{X}$ if

[T 1]: each union of the members of T_O is a member of T_O;
[T 2]: each intersection of a finite number of members of T_O is a member of T_O;
[T 3]: $\mathscr{X}$ belongs to T_O;
[T 4]: the empty set $\emptyset$ belongs to T_O.

Recall that the union of two sets $\mathscr{X}_1$ and $\mathscr{X}_2$ is the set, containing elements of both $\mathscr{X}_1$ and $\mathscr{X}_2$. The intersection of $\mathscr{X}_1$ and $\mathscr{X}_2$ is the set, containing those elements which belong simultaneously to $\mathscr{X}_1$ and $\mathscr{X}_2$. The intersection of two sets may be empty.

The finiteness requirement in point [T 2] is of little relevance for the present discussion.

Let $\mathscr{X} = \{x, y, z, \ldots\}$ be a set. A metric on $\mathscr{X}$ is defined via the requirements:

[M 1]: $d(x, y) \geq 0$ for all $x, y \in \mathscr{X}$;
[M 2]: $d(x, y) = 0$ if and only if $x = y$;
[M 3]: $d(x, y) = d(y, x)$ for all $x, y \in \mathscr{X}$;
[M 4]: $d(x, y) \leq d(x, z) + d(y, z)$ for all $x, y, z \in \mathscr{X}$.

Then $d(x, y)$ is called the distance between the elements x and y.

If a metric is defined on the set $\mathscr{X}$, then $\mathscr{X}$ is called a metric space.

If $\mathscr{X}$ is a metric space, then for each $x \in \mathscr{X}$ and $\varepsilon > 0$ we define the ball-neighbourhood of x as the set of all elements of $\mathscr{X}$ whose distance to x is smaller than ε:

$$\mathscr{U}_\varepsilon(x) = \{y | y \in \mathscr{X}, d(x, y) < \varepsilon\} \, .$$

This ball-neighbourhood is said to have radius ε. Each ball-neighbourhood $\mathscr{U}_\varepsilon(x)$ contains at least one element, namely x itself.

Let us apply these general concepts to graphs. If $G = (\mathscr{V}, \mathscr{E})$ is a connected graph[1], then for any two vertices $u, v \in \mathscr{V}$ the distance $d(u, v)$ is a well-defined quantity (see paragraph 4.1.4) and evidently obeys the above conditions required for a metric.

Consequently, we may speak about ball-neighbourhoods for every vertex. For obvious reasons, only the ball-neighbourhoods with integer radii are to be considered.

For instance, we present the ball-neighbourhoods of radii $\varepsilon = 1, 2, 3, 4$ and 5 of the vertex 1 of the graph G_1.

---

[1] All molecular graphs are connected. Note, however, that as a consequence of the present approach, catenanes, rotaxanes and molecular complexes have to be understood as aggregates composed of several topologically disjoint moieties.

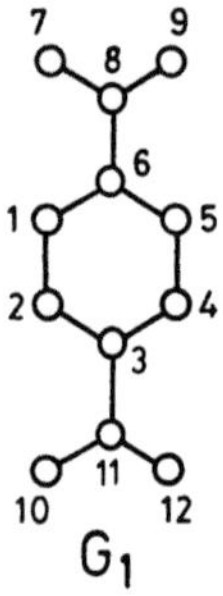

$$G_1$$

$$\mathscr{U}_1(1) = \{1\}$$

$$\mathscr{U}_2(1) = \mathscr{U}_1(1) \cup \{2, 6\} = \{1, 2, 6\}$$

$$\mathscr{U}_3(1) = \mathscr{U}_2(1) \cup \{3, 5, 8\} = \{1, 2, 3, 5, 6, 8\}$$

$$\mathscr{U}_4(1) = \mathscr{U}_3(1) \cup \{4, 7, 9, 11\} = \{1, 2, 3, 4, 5, 6, 7, 8, 9, 11\}$$

$$\mathscr{U}_5(1) = \mathscr{U}_4(1) \cup \{10, 12\} = \{1, 2, \ldots, 12\} = \mathscr{V}(G_1)$$

Evidently, $\mathscr{U}_\varepsilon(1) = \mathscr{V}(G_1)$ for all $\varepsilon \geq 5$.

We define now the neighbourhood $\mathscr{U}$ of an element x of the set $\mathscr{X}$ as any subset of $\mathscr{X}$, such that for some $\varepsilon > 0$, $\mathscr{U}_\varepsilon(x) \subseteq \mathscr{U}$. Let $U(x)$ be the collection of all neighbourhoods of x. A *neighbourhood topology* or a *U-topology* ($U = $ Umgebung) T_U in the set $\mathscr{X}$ is defined by means of the neighbourhoods $U(x)$, $x \in \mathscr{X}$ provided the following conditions are obeyed:

[U 1]: $x \in \mathscr{U}$ for all $\mathscr{U}$ from $U(x)$;

[U 2]: if $\mathscr{U}$ is a member of $U(x)$ and $\mathscr{U} \subseteq \mathscr{U}'$, then $\mathscr{U}'$ is a member of $U(x)$; $\mathscr{X}$ is a member of $U(x)$;

[U 3]: the intersection of two members of $U(x)$ is a member of $U(x)$;

[U 4]: for every $\mathscr{U} \in U(x)$ there exists a $\mathscr{U}'' \in U(x)$, such that $\mathscr{U} \in U(y)$ for all $y \in \mathscr{U}''$.

The statements [U 1]–[U 4] are known as the HAUSDORFF *neighbourhood axioms*. The set $\mathscr{X}$ together with its *U*-topology forms a *topological space*.

If we apply the above definitions to the vertex set $\mathscr{V}(G)$ of a connected graph G, then it is easy to see that the collection $U(v)$ of all neighbourhoods of a vertex v satisfies the HAUSDORFF axioms [U 1]–[U 4]. Hence the set of all $U(v)$, $v \in \mathscr{V}(G)$ defines *a neighbourhood topology in the vertex set of a molecular graph and this is just what we shall call the topology of the corresponding molecule.* The vertex set together with its neighbourhood topology forms the *molecular topological space*.

It should be noted that a *U*-topology, contrary to an *O*-topology does not contain the empty set. Indeed, $U(x)$ cannot contain the empty set since every neighbourhood $\mathscr{U} \in U(x)$ has at least one element, x itself (c.f. axiom [U 1]). On the other hand, the set $U(x)$ amended by the empty set is an *O*-topology, as the reader may easily verify in the case of the vertex set of a molecular graph.

Part B

Chemistry and Graph Theory

Chapter 3

Chemical Graphs

The concept of a molecular graph was introduced in Sect. 2.1. Elements of the mathematical apparatus of graph theory will be given in Chap. 4. The purpose of the present chapter is to review the manifold types of molecular graphs which occur in mathematical investigations in organic chemistry.

The most straightforward graph representation of a molecule has already been mentioned in Sect. 2.1. We shall call it the *complete molecular graph* and illustrate it by two further examples:

The *skeleton graph* represents only those atoms which can be regarded as forming the framework of the molecule. Other atoms, which are usually hydrogens, are disregarded. The skeleton graph is almost universally used in applications of graph theory to saturated and fully conjugated hydrocarbons, since then the neglect of the hydrogen atoms cannot cause any ambiguity.

The following are the skeleton graphs of seven monocycloalkanes C_6H_{12}:

The first of them represents cyclohexane, the second methylcyclopentane, the third 1,1-dimethylcyclobutane, etc.

The skeleton graphs of fully conjugated molecules, not necessarily hydrocarbons, are called HÜCKEL *graphs* because of their role in HÜCKEL molecular orbital theory (see Chap. 5). They correspond, in fact, to the network over which the π-electrons in such molecules are delocalized. It must be immediately pointed out that the use of HÜCKEL graphs goes much beyond the HÜCKEL molecular orbital scheme. Historically, however, they were first used by ERICH HÜCKEL in the early thirties[1].

Here are some examples:

The fact that different conjugated molecules (in our example: pteridine and naphthalene) can have isomorphic HÜCKEL graphs is, of course, rather inconvenient and calls for amendments. The natural solution of this problem is to associate pertinent weights to the vertices and edges of the molecular graph. The theory of such "weighted" graphs is elaborated in Sect. 6.5.

In the study of σ-electronic systems a slightly more complicated graph representation is needed, which is in fact the *line graph of the complete molecular graph*. We present here only an example and refer the interested reader to Sect. 4.4 (for the definition of line graphs) and Chap. 13 (for the role which such graphs play in MO theory).

Some special types of graphs have been used in the topological investigations of benzenoid hydrocarbons. The *inner dual* is obtained by associating a vertex to each hexagon and connecting vertices corresponding to adjacent hexagons. The inner dual of benzo[a]pyrene is given as follows:

---

[1] That the Hamiltonian of the HÜCKEL molecular orbital theory is closely related to the adjacency matrix of a graph was recognized much later, as a result of the work of RUEDENBERG, COULSON, HEILBRONNER and others.

Different benzenoids may have identical inner duals. For instance the inner dual of both chrysene and tetracene is isomorphic to the path with four vertices:

In order to overcome such a redundancy (and thus make the concept of inner dual more suitable for application) BALABAN [72] defined the *characteristic* (or dualist) *graph* of a benzenoid system as the diagram obtained in the same way as the inner dual, but in which the vertices retain their original geometric positions. Hence the characteristic graphs of chrysene and tetracene are different:

It is clear that the characteristic graph (in spite of its name) is not a graph at all. Namely, a vertex in a graph cannot have any "position"; the unique property which a vertex possesses in a graph is its connectedness to other vertices. This, of course, does not mean that we are denying the documented usefulness of characteristic graphs in the theory of polycyclic aromatic hydrocarbons [73].

Two further graphs have been associated with benzenoid hydrocarbons: the GUTMAN *trees* [86, 101] and the CLAR *graphs* [106]. They play a certain role in the aromatic sextet theory of CLAR [13] and in the theory of the sextet polynomial [132]; none of these will be considered in the present book. We show as an example the GUTMAN tree and the CLAR graph of a non-branched cata-condensed benzenoid hydrocarbon as well as the CLAR graph of a peri-condensed molecule:

Note that GUTMAN trees exist only for non-branched cata-condensed benzenoids. In that case the CLAR graph is the line graph of the corresponding GUTMAN tree. For more details along these lines see [86, 101, 106].

The *factor graph* concept has been introduced by JOELA [135]. A factor graph is constructed from a KEKULÉ valence formula of a conjugated molecule by associating a vertex to each double bond and connecting vertices corresponding to double bonds separated by one single bond. The five factor graphs of phenanthrene are given as follows:

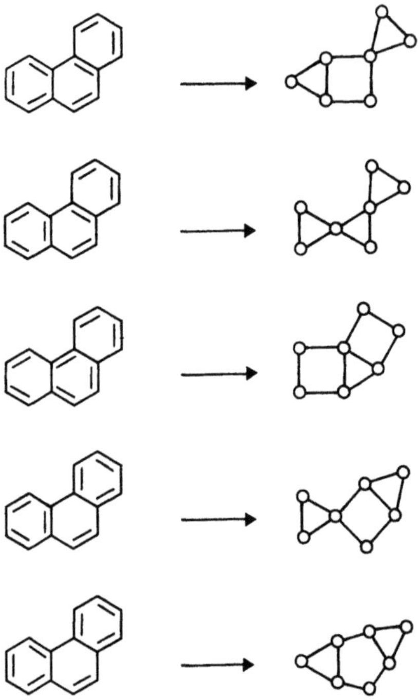

For various applications of factor graphs see [87] and the references cited therein.

The present account of types of molecular graphs used in the chemical literature is not intended to be complete. Rather we wanted to illustrate the many facets of the application of graph theory in modeling chemical phenomena.

A rather important class of chemical graphs has not been mentioned so far. These are the *reaction graphs* in which vertices represent different chemical systems and two vertices are connected if the corresponding species can be chemically inter-converted one into other. Reaction graphs are used both in the study of the kinetics of chemical processes and in classification of chemical reactions. New reaction pathways could be discovered by this means.

Reaction graphs go beyond the scope of the present book. Nowadays they have an extensive literature; the interested reader is referred to [6, 39, 75, 77, 85, 139].

Chapter 4

Fundamentals of Graph Theory

It is not necessary to persuade the reader that graphs are one of the basic mathematical objects with which the present book is concerned. Chap. 2 gave a general conceptual basis for the use of graphs for representing the topology of a molecule. In Chap. 3 we got acquainted with a variety of types of molecular graphs. The present chapter will, finally, provide a precise mathematical characterization of a graph. We shall list here a number of additional graph-theoretical definitions and mention a few basic properties of graphs.

Only those graph-theoretical notions which will be needed later will be introduced here. Hence this chapter can by no means provide a substitute for a text-book of graph theory. Those readers who wish to get a more complete knowledge of graph theory should study one of the many existing books on this subject (e.g. [8, 10, 34, 55, 57, 59, 67]). For first reading the textbook of HARARY [34] is usually recommended.

In addition to the present, more or less, abstract outline of the basic concepts of graph theory, in Chap. 6 a number of chemically important classes of graphs are examined in some detail. Further applications of graph theory are found in the Chaps. 11, 12, 13 and elsewhere.

4.1 The Definition of a Graph

4.1.1 Relations

Let $\mathscr{V}$ be a set of some unspecified elements, $\mathscr{V} = \{v_1, v_2, \ldots, v_n\}$. In the following we shall always assume that $\mathscr{V}$ is a finite set, that is the number of elements of $\mathscr{V}$ is finite.

The set $\mathscr{V} \otimes \mathscr{V}$ consists of all ordered pairs $[v_r, v_s]$ of the elements of $\mathscr{V}$. Any subset $\mathscr{R}$ of $\mathscr{V} \otimes \mathscr{V}$ is called a *relation* on the set $\mathscr{V}$.

For example, $\{[v_2, v_3], [v_2, v_5], [v_3, v_3], [v_4, v_5], [v_5, v_4]\}$ is a relation on the set $\{v_1, v_2, v_3, v_4, v_5\}$.

A relation on a set can be presented diagrammatically if we symbolize the elements of $\mathscr{V}$ by small circles and draw a directed line between two circles whenever the corresponding ordered pair belongs to the relation. For example the diagram of the above relation is:

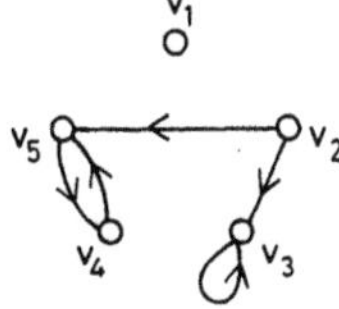

Let $\mathscr{R}$ be a relation on the set $\mathscr{V}$ and let u, v be elements of $\mathscr{V}$. Then $\mathscr{R}$ is said to be *symmetric* if $[v_r, v_s] \in \mathscr{R}$ implies $[v_s, v_r] \in \mathscr{R}$. The relation $\mathscr{R}$ is said to be *reflexive* if for all $v \in \mathscr{V}$, $[v, v] \in \mathscr{R}$. The relation is *antireflexive* if $[v, v] \in \mathscr{R}$ is not true for any $v \in \mathscr{V}$. In other words, $\mathscr{R}$ is antireflexive if

$$[v_r, v_s] \in \mathscr{R} \text{ implies } v_r \neq v_s.$$

For example, $\{[v_2, v_5], [v_4, v_5], [v_5, v_2], [v_5, v_4]\}$ is a symmetric and antireflexive relation on the set $\{v_1, v_2, v_3, v_4, v_5\}$ and has the following diagramatic representation.

If a relation is symmetric and antireflexive then the ordered pairs in it appear in doublets $[v_r, v_s]$ and $[v_s, v_r]$. Instead of such a doublet, one can take an unordered pair (v_r, v_s). Hence a *symmetric and antireflexive relation on a set $\mathscr{V}$ is a set of certain unordered pairs of the elements of $\mathscr{V}$*. In the diagramatical representation, a pair of oppositely directed lines can be substituted by a single undirected line.

The relation given in the previous example can be written also as $\{(v_2, v_5), (v_4, v_5)\}$ and the corresponding diagram as

4.1.2 The First Definition of a Graph

Consider a finite set $\mathscr{V}$. Let $\mathscr{E}$ be a symmetric and antireflexive relation on $\mathscr{V}$, that is let $\mathscr{E}$ be a set of unordered pairs of elements from $\mathscr{V}$. Then *the set $\mathscr{V}$ together with the relation $\mathscr{E}$ forms a graph*. If this graph is denoted by G, then we shall write $G = (\mathscr{V}; \mathscr{E})$.

The last example considered in the previous section: $\mathscr{V} = \{v_1, v_2, v_3, v_4, v_5\}$ and $\mathscr{E} = \{(v_2, v_5), (v_4, v_5)\}$ was, in fact, a graph.

The procedure for the pictorial representation of a graph $G = (\mathscr{V}, \mathscr{E})$ is evident: the elements from $\mathscr{V}$ are visualized by means of small circles and two such circles are joined by a line if the corresponding elements of $\mathscr{V}$ are in the relation $\mathscr{E}$.

The elements of the set $\mathscr{V}$ are called *vertices*. (Some authors call them points or nodes.) The elements of the relation $\mathscr{E}$, considered as unordered pairs, are called *edges*. (Some authors call them lines.)

The present definition of a graph can easily be extended to the case when $\mathscr{E}$ is either not symmetric (then we speak about graphs with directed edges or digraphs)

or not antireflexive (then we speak about graphs with loops) or when $\mathscr{E}$ is neither symmetric nor antireflexive. We shall not be concerned with these types of graphs, although they are rather interesting in both "pure" graph theory and its numerous applications.

A graph defined in the above restricted manner is denoted as *simple* or "schlicht". Hence a simple graph does not posses directed edges and/or edges which start and end at the same vertex (loops). Such graphs do not posses multiple edges either. (Multiple edges are discussed in the subsequent paragraph.)

In the present book we deal almost exclusively with simple graphs. The set of all graphs is denoted by $\mathscr{G}$ whereas the set of all graphs with n vertices by $\mathscr{G}_n$.

4.1.3 The Second Definition of a Graph

Consider two finite sets $\mathscr{V} = \{v_1, v_2, \dots, v_n\}$ and $\mathscr{E} = \{e_1, e_2, \dots, e_m\}$. Let f be a mapping which associates each element of $\mathscr{E}$ to an (unordered) pair of elements of $\mathscr{V}$. Hence for any $e_i \in \mathscr{E}$ there exists a unique pair (v_{i_1}, v_{i_2}), $v_{i_1} \in \mathscr{V}$, $v_{i_2} \in \mathscr{V}$, such that

$$f: e_i \to (v_{i_1}, v_{i_2}) . \tag{1}$$

Then *the two sets $\mathscr{V}$ and $\mathscr{E}$, together with the mapping f form a graph.* This graph can be denoted as $(\mathscr{V}, \mathscr{E}, f)$.

For example, the graph considered in the previous two subsections can be defined by means of $\mathscr{V} = \{v_1, v_2, v_3, v_4, v_5\}$, $\mathscr{E} = \{e_1, e_2\}$ and the mapping f,

$$f: e_1 \to (v_2, v_5) \tag{2}$$
$$e_2 \to (v_4, v_5) .$$

The advantage of this definition is that it can easily be extended to multigraphs, that is graphs with multiple edges. Really, the mapping f need not be a bijection; several elements of $\mathscr{E}$ could be mapped onto the same pair of elements of $\mathscr{V}$.

For example, if $\mathscr{V} = \{v_1, v_2, v_3, v_4\}$, $\mathscr{E} = \{e_1, e_2, e_3, e_4, e_5\}$ and

$$f: e_1 \to (v_1, v_2) \tag{3}$$
$$e_2 \to (v_2, v_3)$$
$$e_3 \to (v_2, v_3)$$
$$e_4 \to (v_2, v_3)$$
$$e_5 \to (v_3, v_4)$$

then we have the following multigraph $(\mathscr{V}, \mathscr{E}, f)$:

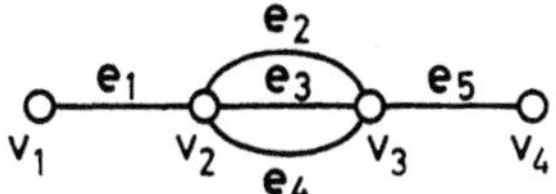

The theory of multigraphs will not be further elaborated in the present book[1]. Nevertheless, the following remark is worth noting. If a unit "weight" is associated to a simple edge, then it is natural to interpret a multiple edge as having weight greater than unity (e.g. 2 and 3 or $\sqrt{2}$ and $\sqrt{3}$ for a double and a triple edge, respectively). Hence, the multigraphs provide a bridge between the simple graphs and the so-called weighted graphs. More details on weighted graphs can be found in Sect. 6.5.

4.1.4 Vertices and Edges

If $G = (\mathscr{V}, \mathscr{E})$ is a graph then the elements of the set $\mathscr{V}$ are called vertices and the elements of the set $\mathscr{E}$ edges. Throughout the entire book the number of vertices will be denoted by n and the number of edges by m. We may symbolize this also as $|\mathscr{V}| = n$, $|\mathscr{E}| = m$.

The fact that $\mathscr{V}$ is the vertex set of the graph G will often be denoted by $\mathscr{V}(G)$.

Two vertices $u \in \mathscr{V}$ and $v \in \mathscr{V}$ are *adjacent* if $(u, v) \in \mathscr{E}$. Then we shall also say that u and v are neighbouring vertices, that they are connected by an edge $e = (u, v)$, that u and v are the end vertices of the edge e and that the edge e is incident to the vertices u and v.

The number of vertices which are adjacent to a given vertex v is called the *degree* (or valency) of this vertex. A vertex of degree zero is called an *isolated vertex*. A vertex of degree one is called a *pendent vertex*.

For example, the graph G_1,

$$G_1$$

has three isolated and four pendent vertices. In addition to this, G_1 has three vertices of degree 2, a vertex of degree 3 and a vertex of degree 5.

If g_v is the degree of the vertex v, then the following identity holds

$$\sum_{v \in \mathscr{V}} g_v = 2m \tag{4}$$

where, of course, m is the number of edges of the graph[2]. The reader may check whether the relation (4) holds for G_1.

A sequence of vertices $v_{i_0}, v_{i_1}, \dots, v_{i_l}$ of a graph, such that $v_{i_{j-1}}$ and v_{i_j} are adjacent $j = 1, 2, \dots, l$ is called a *path* or *elementary path* in the graph G, connecting the vertices v_{i_0} and v_{i_l}. The length of this path is l because the path contains l edges.

---

[1] There exists an obvious analogy between multiple edges in a multigraph and multiple chemical bonds. The above multigraph has been chosen so as to resemble the structural formula of acetylene. Hardly any useful application of this analogy has so far been made in mathematical organic chemistry.

[2] A chemical consequence of this result is that the number of atoms in a molecule having odd valency must be even.

The vertices between which there exists an elementary path are said to belong to the same *component* of the graph. Otherwise they belong to different components.

The number of components of a graph G is denoted by $k = k(G)$.

For example, $k(G_1) = 5$.

A graph for which $k = 1$ is said to be connected.

If G_a, G_b, G_c, ... , G_k are the components of the graph G, we shall write

$$G = G_a \cup G_b \cup G_c \cup ... \cup G_k \, . \tag{5}$$

Concerning the union of graphs the reader should consult Sect. 4.4.

For connected graphs, the *distance* between two vertices u and v is defined as the length of the shortest path between u and v. The distance between the vertices u and v is denoted by $d(u, v)$ and has already been considered in Sect. 2.3.

Since the notion of distance is crucial in the definition of molecular topology (Sect. 2.3) we present the *distance matrix* $D(G_2)$ of the graph G_2. (The entry in the i-th row and j-th column of the distance matrix is the distance between the vertices v_i and v_j.)

$$D(G_2) = \begin{bmatrix} 0 & 1 & 2 & 3 & 2 & 1 \\ 1 & 0 & 1 & 2 & 2 & 1 \\ 2 & 1 & 0 & 1 & 1 & 2 \\ 3 & 2 & 1 & 0 & 1 & 2 \\ 2 & 2 & 1 & 1 & 0 & 1 \\ 1 & 1 & 2 & 2 & 1 & 0 \end{bmatrix}$$

4.1.5 Isomorphic Graphs and Graph Automorphisms

Let G and H be two graphs whose vertices are u_1, u_2, ... , u_n and v_1, v_2, ... , v_n, respectively. Let further π be a permutation[1] of the numbers $1, 2, ... , n$:

$$\pi : \begin{pmatrix} 1 & 2 & ... & n \\ \pi(1) & \pi(2) & ... & \pi(n) \end{pmatrix} \tag{6}$$

Suppose now that a permutation π can be found, such that for $i = 1, 2, ... , n$ and $j = 1, 2, ... , n$ the vertices u_i and u_j are adjacent in G if and only if the vertices $v_{\pi(i)}$ and $v_{\pi(j)}$ are adjacent in H. Then the two graphs are said to be *isomorphic*, $G \simeq H$.

In that case the permutation π represents an isomorphic mapping of the vertex set of G onto the vertex set of H; an *isomorphic mapping* of the vertices always conserves the adjacency relations. The vertex u_i is called the origin and the vertex $v_{\pi(i)}$ the image of the mapping.

Isomorphic graphs are essentially one and the same mathematical object. They differ only in the way in which their vertices are labeled.

---

[1] For the definition of a permutation see Appendix 2. Basic properties of permutations are outlined in Chap. 9.

An isomorphic mapping of the vertices of a given graph onto themselves (which also preserves the adjacency relation) is called an *automorphism* of the graph. Evidently, each graph possesses a trivial automorphism, the so-called *identity automorphism*:

$$\pi_0 : \begin{pmatrix} 1 & 2 & \dots & n \\ 1 & 2 & \dots & n \end{pmatrix} \tag{7}$$

Some graphs may also possess *non-identical automorphisms*. For example, in the case of G_2

$$\pi_1 : \begin{pmatrix} 1 & 2 & 3 & 4 & 5 & 6 \\ 1 & 6 & 5 & 4 & 3 & 2 \end{pmatrix} \tag{8}$$

is an automorphism.

The set of all automorphisms of a graph forms a group, the so-called *automorphism group*. This matter is further elaborated in Chap. 9.

A number $I(G)$ which can be associated with a graph G in a certain way and which has the same value for all graphs isomorphic to G is called a *graph invariant*. Consequently, graph invariants are quantities independent of the labeling of the vertices of a graph.

We mention in passing that determining whether two given graphs are isomorphic or not is by no means an easy task. A reader interested in testing his imagination may try to decide which two (if any) of the graphs G_3, G_4 and G_5 are isomorphic.

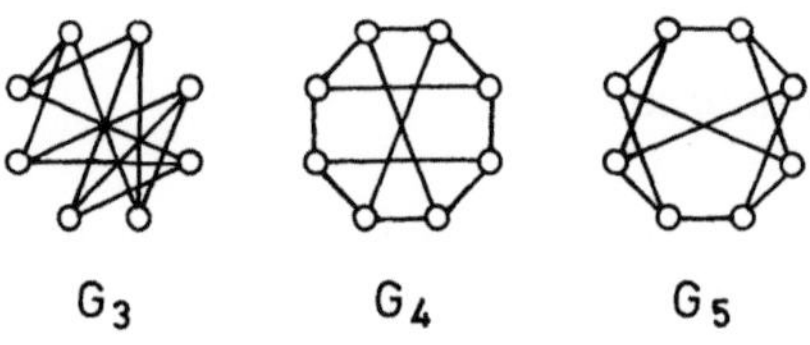

$$G_3 \qquad\qquad G_4 \qquad\qquad G_5$$

4.1.6 Special Graphs

If $\mathscr{E}$ is an empty set, $\mathscr{E} = \emptyset$, then the graph $G = (\mathscr{V}, \mathscr{E})$ consists of n isolated vertices. This is the *null graph*. If $\mathscr{E}$ contains all possible pairs of elements of $\mathscr{V}$, then the corresponding graph is called *complete* and is denoted by K_n. The first five complete graphs are given as follows:

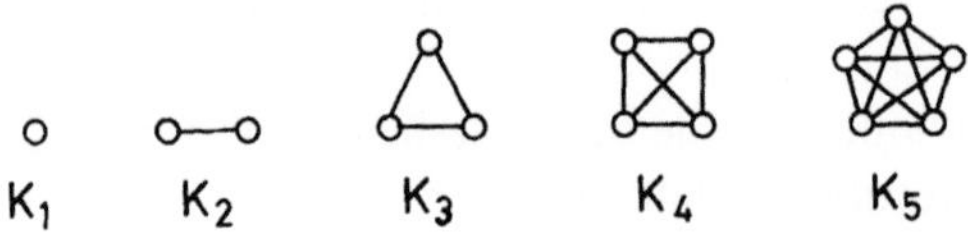

$$K_1 \qquad K_2 \qquad K_3 \qquad K_4 \qquad K_5$$

Evidently, all vertices in K_n are adjacent and the distance between any two of them is unity.

A graph in which all vertices have equal degree, say g, is a *regular graph* of degree g. According to this definition K_n is a regular graph of degree $n - 1$. In chemical applications the connected regular graph of degree 2 plays an outstanding role. Its name is the *cycle* (or circuit) and will be denoted by C_n. The first five cycles are given as follows:

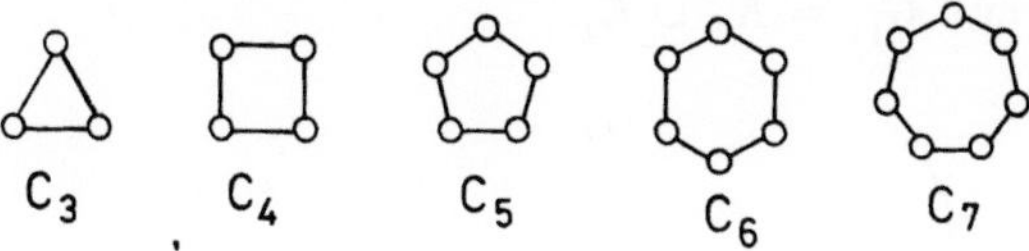

A connected graph which does not contain cycles is called a *tree*. A detailed exposition of the theory of trees is given in Sect. 6.1.

4.2 Subgraphs

If $G = (\mathscr{V}, \mathscr{E})$ is a graph, $\mathscr{V}^*$ is a subset of $\mathscr{V}$ and $\mathscr{E}^*$ both a subset of $\mathscr{E}$ and a symmetric and antireflexive relation on $\mathscr{V}^*$, then $G^* = (\mathscr{V}^*, \mathscr{E}^*)$ is a *subgraph* of the graph G. In other words, a subgraph is obtained by deleting certain vertices and certain edges from the graph. The deletion of a vertex implies also the deletion of all the edges which are incident to this vertex.

G_6, G_7 and G_8 are subgraphs of G_2. Note that G_6 is obtained by deletion of edges, G_7 by deletion of vertices (and the incident edges) while G_8 is obtained by deletion of both vertices and edges.

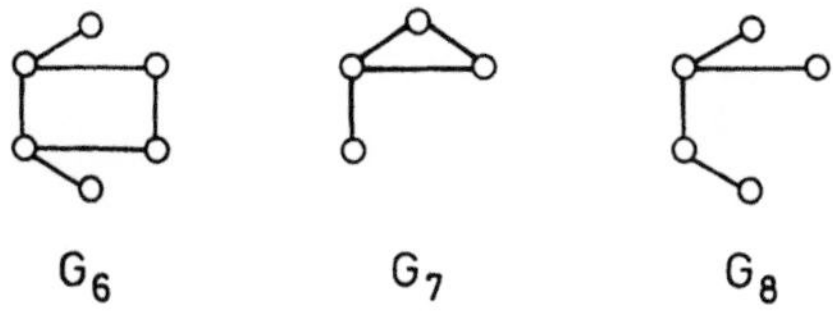

If v is a vertex of the graph G, then $G - v$ will denote the subgraph obtained by deletion of v from G. Similarly $G - e$ is the subgraph obtained by deletion of the edge e. In general, if H is a subgraph of G, $G - H$ will symbolize the deletion of all vertices (and, of course, all incident edges) of H from G.

If H is a subgraph of the graph G we shall say that H is contained in G. In particular, if C_k is contained in G, we say that the graph G contains a k-membered cycle and G is a *cyclic graph*. If G does not contain cycles, then G is *acyclic* (also called a forest). As already mentioned, a connected acyclic graph is a *tree*.

A cycle with n vertices, contained in a graph G with n vertices is called a *Hamiltonian cycle* of G. A tree with n vertices, contained in a graph G with n vertices is a *spanning tree* of G. Only connected graphs have spanning trees.

Two special types of subgraphs play a distinguished role in the chemical applications of graph theory. These are the SACHS graphs and the matchings. The two subsequent paragraphs elaborate this matter in some detail.

4.2.1 Sachs Graphs

A SACHS *graph* is a graph whose components are cycles and/or complete graphs with two vertices, K_2.

For example, G_9, G_{10} and G_{11} are SACHS graphs contained in G_2. We can write $G_9 = C_3 \cup C_3$, $G_{10} = C_3 \cup K_2$ and $G_{11} = K_2 \cup K_2$.

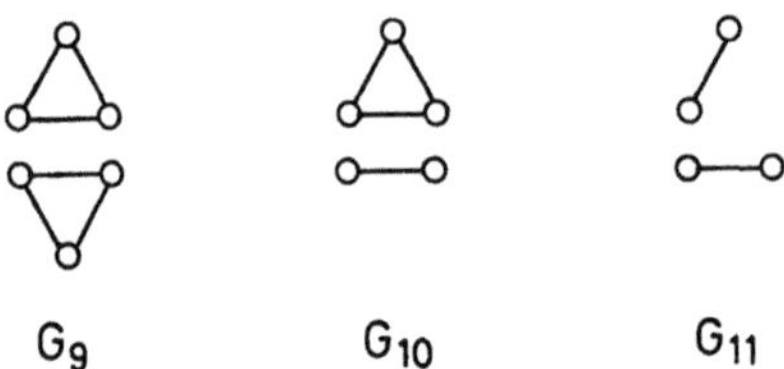

Note that all components of G_9 are cycles, all components of G_{11} are K_2 graphs whereas G_{10} is composed of cycles and K_2 graphs. In the above example all three SACHS graphs have two components. The number of cycles is 2, 1 and 0, respectively; the number of vertices is 6, 5 and 4, respectively.

Let S be a SACHS graph. Then $k(S)$, $c(S)$ and $n(S)$ will denote the number of components, cycles and vertices of S.

For example, consider the following four SACHS graphs S_1, S_2, S_3 and S_4:

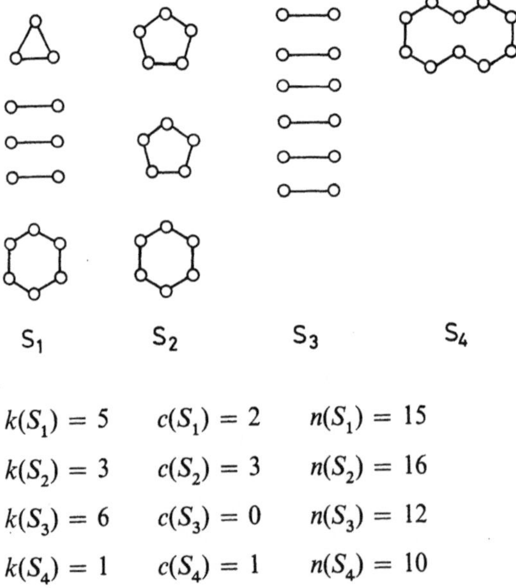

$$k(S_1) = 5 \qquad c(S_1) = 2 \qquad n(S_1) = 15$$
$$k(S_2) = 3 \qquad c(S_2) = 3 \qquad n(S_2) = 16$$
$$k(S_3) = 6 \qquad c(S_3) = 0 \qquad n(S_3) = 12$$
$$k(S_4) = 1 \qquad c(S_4) = 1 \qquad n(S_4) = 10$$

The set of all SACHS graphs with i vertices, which are contained in a graph G, is denoted by $\mathscr{S}_i(G)$ or, when there is no danger of misunderstanding, simply by $\mathscr{S}_i$.

The knowledge of the sets $\mathscr{S}_1, \mathscr{S}_2, \ldots, \mathscr{S}_n$ is essential for the application of the SACHS theorem, which will be extensively discussed later on. Here we want to point out that the finding of all SACHS graphs of a given graph is an extremely laborious and error prone combinatorial task.

Figure 4.1 presents, as an instructive example, the set of the 6-vertex SACHS graphs of the benzocyclobutadiene graph G_{12}.

G_{12}

The following general properties of the sets $\mathscr{S}_i$ are immediate consequences of the definition of SACHS graphs.

1. Since a cycle has at least three vertices and K_2 has two vertices, the SACHS graphs have at least two vertices. Therefore, the set $\mathscr{S}_1$ is always empty, $\mathscr{S}_1(G) = \emptyset$ for all graphs G.

2. The set $\mathscr{S}_2$ contains only SACHS graphs with a unique component, namely K_2. The number of elements in $\mathscr{S}_2(G)$ is equal to the number of edges of the graph G.

3. The set $\mathscr{S}_3$ contains only SACHS graphs with a unique component, namely C_3. The number of elements in $\mathscr{S}_3(G)$ is equal to the number of triangles in the graph G.

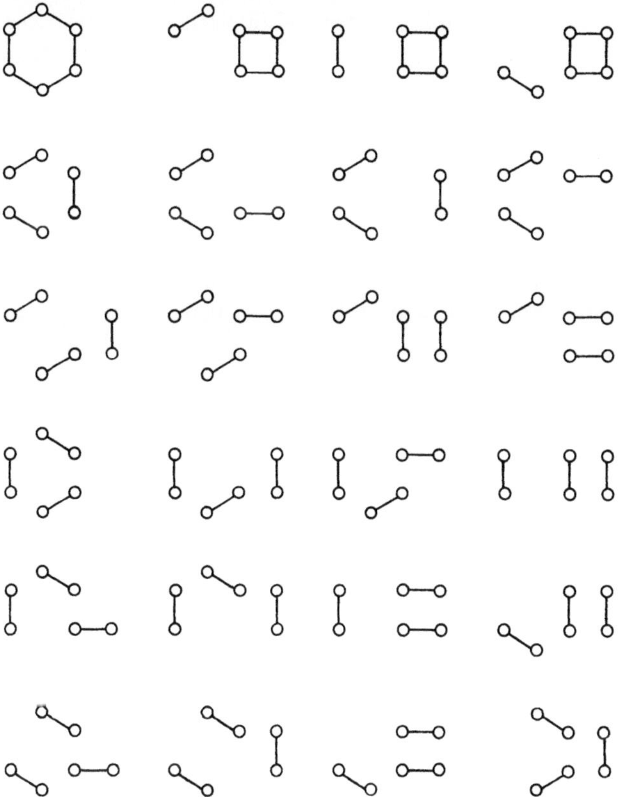

Fig. 4.1. The set of all SACHS graphs with 6 vertices of the benzocyclobutadiene graph. This example illustrates how difficult is the problem of finding these subgraphs even in the case of molecular graphs with small number of vertices

4. Since the number of vertices of K_2 is even, every SACHS graph with odd number of vertices must contain at least one odd-membered cycle. Consequently, if the graph G does not contain odd-membered cycles, $\mathscr{S}_i(G) = \emptyset$ for all odd values of i. Note that graphs without odd-membered cycles, are the so-called bipartite graphs, discussed in detail in Sect. 6.3.

4.2.2 Matchings

Two edges of a graph are said to be independent if they are not incident to a common vertex. A k-matching of the graph G is a selection of k mutually independent edges. It is immediately clear that every k-matching corresponds to a subgraph of G, containing k copies of K_2.

Bearing in mind the definition of a SACHS graph we conclude that every SACHS graph with k components, all of which are K_2-graphs (hence, none of which is a cycle) is in a one-to-one correspondence to a k-matching. The previously given SACHS graphs G_{11} and S_3 represent a 2- and a 6-matching, respectively. Similarly, there are 20 distinct 3-matchings of G_{12} and they are presented by the last 20 SACHS graphs in Fig. 4.1.

The number of k-matchings of the graph G is denoted by $m(G, k)$.

In particular, $m(G, 1)$ is equal to the number of edges of G. For all graphs G we define $m(G, 0) = 1$. Since every k-matching covers $2k$ vertices of a graph, there cannot exist k-matchings with $k > n/2$. Hence $m(G, k) = 0$ for $k > n/2$.

For example, $m(G_{12}, 3) = 20$. We shall demonstrate in a while that $m(G_{12}, 4) = 3$. In addition to this, $m(G_{12}, 1) =$ number of edges of $G_{12} = 9$.

If a graph G possesses n vertices and n is even, then the $(n/2)$-matchings of G are called *perfect matchings* (or linear factors) of G. An interesting observation, which will not be further elaborated in this book, is that every perfect matching of a HÜCKEL graph is in a one-to-one correspondence with a KEKULÉ structure. One example should suffice:

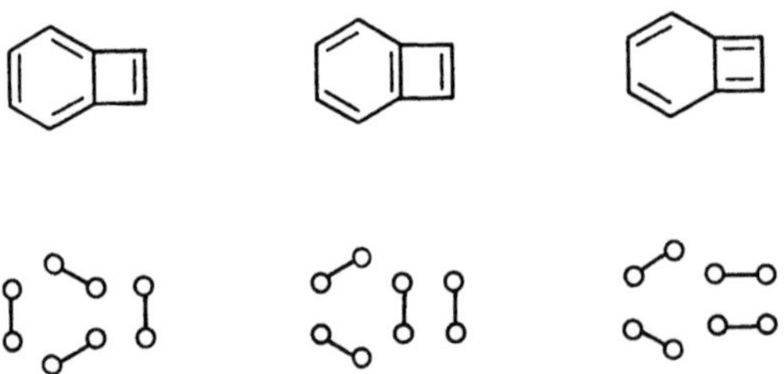

As a consequence of this, if n is even then $m(G, n/2)$ is equal to the number of KEKULÉ structures of the corresponding conjugated molecule.

We prove now the following important property of the numbers $m(G, k)$.

Lemma 4.1. *If u and v are adjacent vertices of G and e is the edge connecting u and v, then for $k \geq 1$*

$$m(G, k) = m(G - e, k) + m(G - u - v, k - 1) \, . \tag{9}$$

Proof. All the k-matchings of G can be divided into two groups: those which contain the edge e and those which do not contain e. The way in which we can select k independent edges from G, so that e is not among the selected edges is evidently equal to the total number of selections of k independent edges in $G - e$. This is just $m(G - e, k)$.

If the edge e is selected, then we have to find additional $k - 1$ independent edges of G, which, consequently, must not be incident to the vertices u and v. The number of such selections is $m(G - u - v, k - 1)$. $\square$

Lemma 4.1 has an important special case, namely when the vertex v is pendent. Then Eq. (9) becomes

$$m(G, k) = m(G - v, k) + m(G - u - v, k - 1) . \tag{10}$$

The *matching polynomial* of the graph G is defined as

$$\alpha(G) = \alpha(G, x) = \sum_{k=0}^{[n/2]} (-1)^k \, m(G, k) \, x^{n-2k} \tag{11}$$

and plays an important role in many chemical applications. A complete theory of the matching polynomial will not be outlined in the present book and the readers are referred to the reviews [23, 27]. Some properties of the matching polynomial are given in Theorem 11.5.

4.3 Graph Spectral Theory

The theory of graph spectra is extensively used throughout the present book, especially in Chaps. 5, 6, 12 and 13. We give here the necessary definitions and a few basic theorems. Their elaboration and application follows in the subsequent parts of the book. For those who wish to become perfectly familiar with graph spectral theory we recommend the reading of the monograph of CVETKOVIĆ, DOOB and SACHS [16]. Graph spectral theory can also be found in a considerable number of "chemistry oriented" books and reviews [3, 18, 24, 30, 31, 64, 65].

4.3.1 The Adjacency Matrix

The information about the adjacency relation for the vertices $v_1, v_2, \ldots, v_n$ of a graph G can be converted into a matrix form using the following natural definition.

The *adjacency matrix* $A = A(G)$ of the graph G is a square matrix of order n whose entry in the i-th row and j-th column is defined as

$$a_{ij} = \begin{cases} 1 & \text{if the vertices } v_i \text{ and } v_j \text{ are adjacent} \\ 0 & \text{otherwise} . \end{cases} \tag{12}$$

"Otherwise" means that either the vertices v_i and v_j are not adjacent or $i = j$.

The construction of the adjacency matrix depends on the labeling of the vertices of the graph. Hence $A(G)$ is not a graph invariant.

The adjacency matrix of the graph G_2 with vertices labeled as before is

$$A(G_2) = \begin{bmatrix} 0 & 1 & 0 & 0 & 0 & 1 \\ 1 & 0 & 1 & 0 & 0 & 1 \\ 0 & 1 & 0 & 1 & 0 & 0 \\ 0 & 0 & 1 & 0 & 1 & 0 \\ 0 & 0 & 0 & 1 & 0 & 1 \\ 1 & 1 & 0 & 0 & 1 & 0 \end{bmatrix}. \tag{13}$$

It is evident that $A(G)$ is a symmetric matrix with zero diagonal. If the graph G is composed of two components G_a and G_b, $G = G_a \cup G_b$, then $A(G)$ has the following block form

$$A(G) = \begin{bmatrix} A(G_a) & \mathbf{0} \\ \mathbf{0} & A(G_b) \end{bmatrix} \tag{14}$$

where $\mathbf{0}$ is the zero matrix of pertinent order.

4.3.2 The Spectrum of a Graph

The eigenvalue-eigenvector problem of the adjacency matrix of a graph has attracted the attention of graph theoreticians over a considerable period of time (see [16] for further details, references and historical remarks). The HÜCKEL molecular orbital theory has been developed by chemists independently and it was not immediately recognized that the mathematical basis of the HMO theory is just the eigenvalue-eigenvector problem of the HÜCKEL graph (see Chap. 5). After this had been made, the graph spectral theory found numerous chemical applications.

If A is the adjacency matrix of the graph G and there exists a vector C and a number λ, such that

$$AC = \lambda C \tag{15}$$

then C is an *eigenvector* and λ an *eigenvalue* of the graph G.

We assume that the readers are familiar with the eigenvector-eigenvalue theory of real symmetric matrices (see Appendix 3). The following statements about graph eigenvectors and graph eigenvalues are straightforward specializations of statements known for real symmetric matrices.

1. There are n linearly independent graph eigenvectors, $C_1, C_2, \ldots, C_n$. They can always be chosen so as to be orthonormal:

$$C_i^\dagger C_j = \delta_{ij} \tag{16}$$

with δ_{ij} denoting the KRONECKER delta ($\delta_{ij} = 1$ if $i = j$ and $\delta_{ij} = 0$ if $i \neq j$).

2. The graph eigenvalues are real numbers. They will be labeled in non-increasing order as

$$\lambda_1 \geqq \lambda_2 \geqq \ldots \geqq \lambda_n \tag{17}$$

and, of course, so that the eigenvalue λ_i corresponds to the eigenvector C_i. Some graph eigenvalues may be equal, then they correspond to *degenerate* eigenvectors; we shall also say that these eigenvalues are degenerate.

The n graph eigenvalues $\lambda_1, \lambda_2, \ldots, \lambda_n$ form the *spectrum* of the graph. The spectrum is a graph invariant. If two nonisomorphic graphs have the same spectrum, then they are said to be *cospectral* (or *isospectral*). Cospectral graphs occur very frequently and it is easy to construct them. The two matching equivalent trees given in paragraph 6.1.4 are examples of cospectral graphs.

The *characteristic polynomial* of the graph is the characteristic polynomial of its adjacency matrix:

$$\varphi(G) = \varphi(G, x) = \det (xI - A) . \tag{18}$$

We now point out three elementary, and in the later text often used, properties of the characteristic polynomial of the graph.

3. Because of (14)

$$\varphi(G_a \cup G_b) = \varphi(G_a) \cdot \varphi(G_b) . \tag{19}$$

4. A graph eigenvalue is a zero of the characteristic polynomial, $\varphi(G, \lambda_i) = 0$ for all i. According to the definition (18), $\varphi(G)$ is a polynomial of degree n, and can be factorized as

$$\varphi(G, x) = \sum_{i=1}^{n} (x - \lambda_i) . \tag{20}$$

5. The characteristic polynomial is a graph invariant.

4.3.3 The Sachs Theorem

We may write the characteristic polynomial of a graph G in the coefficient form

$$\varphi(G, x) = \sum_{i=0}^{n} a_i x^{n-i} . \tag{21}$$

Since the coefficients $a_0, a_1, a_2, \ldots, a_n$ are graph invariants, one can pose the question: *How does the structure of the graph determine the coefficients of the characteristic polynomial?*

Investigations along these lines have a long history both in mathematics and in chemistry and a surprisingly large number of authors has independently reached similar results. Hence the SACHS theorem was discovered by several scientists and the essential parts of it were already known when HORST SACHS published his paper [171].

Nevertheless the theorem relating the coefficients of the characteristic polynomial with the structure of the graph will be associated with HORST SACHS' name in the following.

A full account of the controversies concerning the discovery of the SACHS theorem can be found in [16].

The concept of SACHS graphs has been defined previously. Let $\mathscr{S}_i$ be the set of all SACHS graphs with i vertices, which are contained in the graph G. Let $k(S)$ and $c(S)$ denote the number of components and cycles in a SACHS graph S.

Theorem 4.2. (The SACHS Theorem). For $i \geqq 1$

$$a_i = \sum_{S \in \mathscr{S}_i} (-1)^{k(S)} \, 2^{c(S)} \tag{22}$$

where the summation goes over all elements of the set $\mathscr{S}_i$. In addition, $a_0 = 1$.

As already mentioned, it is quite difficult to construct the sets $\mathscr{S}_i(G)$ even for graphs of moderate size. Therefore the SACHS theorem (in spite of the claims which sometimes appear in the chemical literature) is only of a rather limited applicability for the calculation of the characteristic polynomial of the graphs.

To illustrate this point we shall determine $a_6(G_{12})$, using the SACHS graphs from Fig. 4.1.

$$a_6(G_{12}) = (-1)^1 \, 2^1 + (-1)^2 \, 2^1 + (-1)^2 \, 2^1 + (-1)^2 \, 2^1 + (-1)^3 \, 2^0$$
$$+ (-1)^3 \, 2^0 + (-1)^3 \, 2^0 + \dots + (-1)^3 \, 2^0 = (-1)^1 \, 2^1 + 3[(-1)^2 \, 2^1]$$
$$+ 20[(-1)^3 \, 2^0] = -2 + 3 \cdot 2 - 20 \cdot 1 = -16 \,. \tag{23}$$

Further exercises of this kind can be found elsewhere [64, 65].

The great value of the SACHS theorem becomes evident when generally valid statements about the coefficients of the characteristic polynomial of a graph are needed [94]. Several sophisticated applications of the SACHS theorem will be given in the following chapters. Here we shall work out only a few elementary results, referring to the properties 1–3 of the sets $\mathscr{S}_i$, outlined in paragraph 4.2.1.

1. Since $\mathscr{S}_1 = \emptyset$, for all graphs (without loops) $a_1 = 0$. As an immediate consequence of this,

$$\sum_{i=1}^{n} \lambda_i = 0 \,. \tag{24}$$

The result that the sum of the eigenvalues of a simple graph is zero follows also from the fact that the adjacency matrix has a zero diagonal, and hence its trace is zero.

2. For all $S \in \mathscr{S}_2$, $k(S) = 1$ and $c(S) = 0$ and therefore

$$a_2 = -(\text{number of edges of } G) = -m \,. \tag{25}$$

Using the relation between a_2 and the graph eigenvalues:

$$a_2 = \sum_{i<j}^{n} \lambda_i \lambda_j \tag{26}$$

and applying (24) one attains to the important identity

$$\sum_{i=1}^{n} \lambda_i^2 = 2m . \tag{27}$$

3. For all $S \in \mathscr{S}_3$, $k(S) = 1$ and $c(S) = 1$ and therefore

$$a_3 = -2 \cdot (\text{number of triangles in } G) . \tag{28}$$

Some additional consequences of the SACHS theorem are also worth mentioning. Let the graph G contain r cycles: $Z_1, Z_2, \dots , Z_r$. Let $G - Z_a$ denote the subgraph obtained by deletion of all the vertices of Z_a from G.

For instance, the benzocyclobutadiene graph G_{12} contains three cycles, Z_1, Z_2 and Z_3:

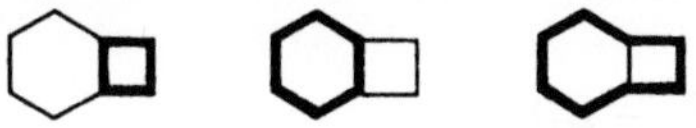

$$Z_1 \qquad Z_2 \qquad Z_3$$

The corresponding subgraphs $G_{12} - Z_1$ and $G_{12} - Z_2$ are given as follows

$$G_{12}\text{-}Z_1 \qquad\qquad G_{12}\text{-}Z_2$$

whereas $G_{12} - Z_3$ is the "graph" $\emptyset$ without vertices.

4. Let e be an edge of the graph G, containing the vertices u and v. Then

$$\varphi(G) = \varphi(G - e) - \varphi(G - u - v) - 2 \sum_{Z} \varphi(G - Z) \tag{29}$$

with the summation going over all cycles Z which contain the edge e. If $G - Z = \emptyset$ then $\varphi(G - Z) = 1$.

If the edge e does not belong to any cycle, that is if e is a bridge, then Eq. (29) reduces to the so-called HEILBRONNER formula:

$$\varphi(G) = \varphi(G - e) - \varphi(G - u - v) . \tag{30}$$

5. Let u be a vertex of the graph G, being adjacent to the vertices $v_1, v_2, \dots , v_g$. Then

$$(\varphi G) = x\varphi(G - u) - \sum_{i=1}^{g} \varphi(G - u - v_i) - 2 \sum_{Z} \varphi(G - Z) \tag{31}$$

where now the second summation on the right-hand side embraces all cycles Z which contain the vertex u.

If the vertex u does not belong to any cycle, that is if u is a cutpoint, then Eq. (31) reduces to another formula associated with the name of EDGAR HEILBRONNER:

$$\varphi(G) = x\varphi(G - u) - \sum_{i=1}^{g} \varphi(G - u - v_i) \,. \tag{32}$$

As a matter of fact, in HEILBRONNER's paper [126] the most general recursion relation of this type, namely Eq. (29) can be found.

The characteristic polynomial, given by Eq. (18), and the matching polynomial, given by Eq. (11), seem to have very little in common. However the following relations between them exist [119]:

6. Let $Z_1, Z_2, \dots, Z_r$ be the cycles of the graph G. If Z_a and Z_b do not possess common vertices, then by definition $G - Z_a - Z_b = (G - Z_a) - Z_b$. If, however, Z_a and Z_b possess common vertices, then the subgraph $G - Z_a - Z_b$ is undefined. In that case one has to set $\varphi(G - Z_a - Z_b) = 0$ and $\alpha(G - Z_a - Z_b) = 0$. The subgraphs $G - Z_a - Z_b - Z_c, G - Z_a - Z_b - Z_c - Z_d$ etc. and their characteristic and matching polynomials are determined analogously. Let, in addition, $\alpha(\emptyset) = 1$. Bearing these definitions and conventions in mind, we have [119]

$$\varphi(G) = \alpha(G) - 2\sum_a \alpha(G - Z_a) + 4\sum_{a<b} \alpha(G - Z_a - Z_b)$$
$$- 8\sum_{a<b<c} \alpha(G - Z_a - Z_b - Z_c) + \dots \tag{33}$$

$$\alpha(G) = \varphi(G) + 2\sum_a \varphi(G - Z_a) + 4\sum_{a<b} \varphi(G - Z_a - Z_b)$$
$$+ 8\sum_{a<b<c} \varphi(G - Z_a - Z_b - Z_c) + \dots \,. \tag{34}$$

4.3.4 The μ-Polynomial

As we have seen in point 6 of the previous paragraph, the characteristic and the matching polynomials of a graph are intimately related. Therefore it is of some interest to seek for some more general polynomials which would, as special cases, yield $\varphi(G)$ and $\alpha(G)$. It is possible to generalize $\alpha(G)$ and $\varphi(G)$ in several different ways. In the present section we shall point out one such attempt, which has found noteworthy applications in the theory of cyclic conjugation [110, 112, 119].

In order to introduce the μ-polynomial we associate a variable weight t_a to the cycle Z_a of the graph G, $a = 1, 2, \dots, r$. The r-dimensional vector $(t_1, t_2, \dots, t_r)$ will be denoted by T.

Let now S be a SACHS graph of the graph G and let the cycles $Z_{a_1}, Z_{a_2}, \dots, Z_{a_c}$ of G be components of S. Then we associate the weight $t(S) = t_{a_1} t_{a_2} \dots t_{a_c}$ to the SACHS graph S. If S is acyclic (that is $c = 0$), then $t(S) = 1$.

In full analogy with the SACHS theorem we define the numbers f_i, $i \geqq 1$ as

$$f_i = \sum_{S \in \mathscr{S}_i} (-1)^{k(S)} 2^{c(S)} t(S) \tag{35}$$

where the summation goes over all SACHS graphs contained in the set $\mathscr{S}_i$. These numbers are interpreted as the coefficients of the μ-polynomial. In other words, the μ-polynomial of the graph G is defined as

$$\mu(G, T, x) = x^n + f_1 x^{n-1} + f_2 x^{n-2} + \dots + f_{n-1} x + f_n . \tag{36}$$

It can be shown that $\mu(G, T)$ satisfies the following recursion relations [119]:

$$\mu(G, T) = \alpha(G) - 2\sum_a t_a \alpha(G - Z_a) + 4 \sum_{a<b} t_a t_b \alpha(G - Z_a - Z_b)$$
$$- 8 \sum_{a<b<c} t_a t_b t_c \alpha(G - Z_a - Z_b - Z_c) + \dots \tag{37}$$

$$\mu(G, T) = \varphi(G) + 2\sum_a (1 - t_a) \varphi(G - Z_a) + 4 \sum_{a<b} (1 - t_a)(1 - t_b) \varphi(G - Z_a - Z_b)$$
$$+ 8 \sum_{a<b<c} (1 - t_a)(1 - t_b)(1 - t_c) \varphi(G - Z_a - Z_b - Z_c) + \dots . \tag{38}$$

From the above formulas is easy to see that for $T = (1, 1, \dots, 1)$ the μ-polynomial reduces to the characteristic polynomial whereas for $T = (0, 0, \dots, 0)$ the μ-polynomial gives as a special case the matching polynomial.

For further details of the mathematical properties and chemical applications of the μ-polynomial the interested reader should consult the paper [119].

4.4 Graph Operations

There are many graph operations known and extensively studied in the mathematical literature. We mention here only those which we shall need later on.

The *complement* of a graph G is the graph $\bar{G}$, which contains the same vertices as G and exactly those edges which are not contained in G.

For example $\bar{G}_{13} = G_{14}$ and $\bar{G}_{14} = G_{13}$:

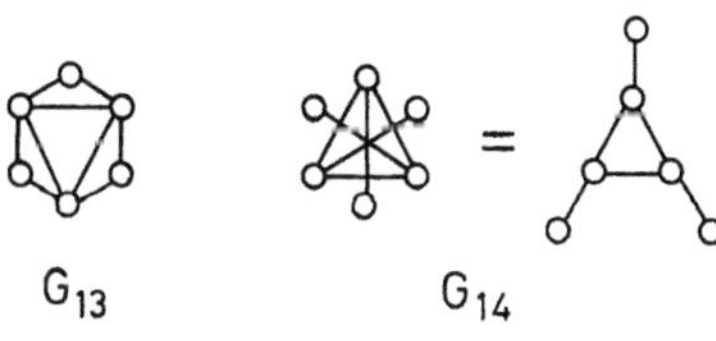

G_{13} $\qquad\qquad$ G_{14}

The complement of the complement of a graph is the graph itself: $\bar{\bar{G}} = G$. The complement of the complete graph K_n is the graph with n vertices and without edges (the so-called null graph).

The *line graph* $L(G)$ of the graph G is constructed so that every vertex of $L(G)$ represents an edge of G. Two vertices of $L(G)$ are adjacent if the corresponding edges of G are incident to a common vertex.

For example the line graph of G_{15} is G_{16}, $L(G_{15}) = G_{16}$:

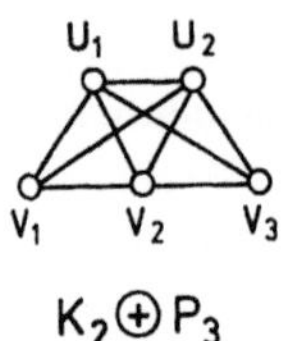

G_{15} G_{16}

The reader should convince himself that $L(C_n) = C_n$.

Let $G_a = (\mathcal{V}_a, \mathcal{E}_a)$ and $G_b = (\mathcal{V}_b, \mathcal{E}_b)$ be two graphs and $\mathcal{V}_a = \{u_1, u_2, \ldots, u_{n_a}\}$, $\mathcal{V}_b = \{v_1, v_2, \ldots, v_{n_b}\}$. Suppose further that the graphs G_a and G_b have disjoint vertex sets, i.e. that the intersection of $\mathcal{V}_a$ and $\mathcal{V}_b$ is empty: $\mathcal{V}_a \cap \mathcal{V}_b = \emptyset$.

The *union* of the graphs G_a and G_b is the graph $G_a \cup G_b$ whose vertex set is $\mathcal{V}_a \cup \mathcal{V}_b$ and whose edge set is $\mathcal{E}_a \cup \mathcal{E}_b$. If $G = G_a \cup G_b$ then one usually says that G_a and G_b are the components of G. We have seen many examples for the union of graphs in the previous parts of this chapter.

The *compound* of the graphs G_a and G_b is the graph $G_a \oplus G_b$ whose vertex set is $\mathcal{V}_a \cup \mathcal{V}_b$ and whose edge set contains the edges of G_a, the edges of G_b and all the edges between the vertices of G_a and G_b.

As an example we present the graph $K_2 \oplus P_3$. Let the vertices of K_2 and P_3 be labeled as indicated.

u_1
u_2 v_1 v_2 v_3

K_2 P_3

Then, $K_2 \oplus P_3$ is the following graph with $2 + 3 = 5$ vertices:

u_1 u_2

v_1 v_2 v_3

$K_2 \oplus P_3$

It is clear from the above definition that $G_a \oplus G_b = G_b \oplus G_a$. It can be shown that

$$G_a \oplus G_b = \overline{\overline{G}_a \cup \overline{G}_b} . \tag{39}$$

In particular, the compound of $\overline{K}_a$ (null graph with a vertices) and $\overline{K}_b$ (null graph with b vertices) is the complete bipartite graph on $a + b$ vertices: $\overline{K}_a \oplus \overline{K}_b = K_{a,b}$. (The notion of a bipartite graph is explained in the paragraph 6.3.1, where examples of complete bipartite graphs also can be found.)

The *composition* of the graphs G_a and G_b, denoted as $G_a[G_b]$ is a graph whose vertex set is $\mathcal{V}_a \otimes \mathcal{V}_b$. Two vertices $[u_i, v_j]$ and $[u_k, v_l]$ of $G_a[G_b]$ are adjacent if either

u_i is adjacent to u_k in G_a

or

$u_i = u_k$ and v_j is adjacent to v_l in G_b .

This relatively complicated definition would become more transparent if the reader would work out a few examples. We give here two simple ones.

Let the vertices of K_2 and P_3 be labeled as before.

Then $K_2[P_3]$ is the following graph with $2 \times 3 = 6$ vertices:

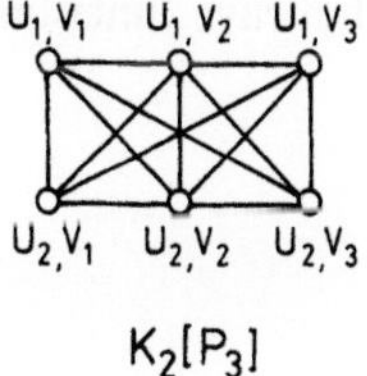

The composition is not a commutative operation, that is $G_a[G_b]$ and $G_b[G_a]$ need not be isomorphic. This is seen from the following example:

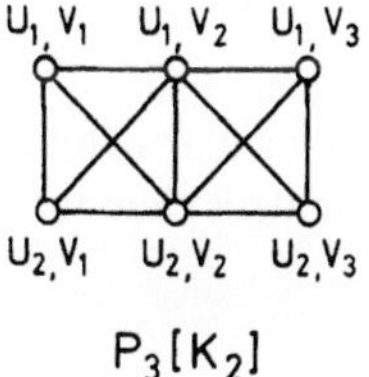

The union, compound and composition of graphs are important in the study of the automorphism groups, which are discussed in some detail in Chap. 9. Here we present only one relevant result of this type.

Theorem 4.3. *The automorphism group of the composition of the graphs G_a and G_b is equal to the wreath product of the automorphism groups of G_a and G_b.*

Chapter 5

Graph Theory and Molecular Orbitals

In the present chapter, as well as throughout the entire book, we assume that the reader knows the basic facts about the HÜCKEL molecular orbital (HMO) theory [35, 51, 62]. Hence HMO theory is an approximate quantum-mechanical approach to the description of the π-electrons in unsaturated conjugated molecules. The wave function for a π-electron is presented in the LCAO form

$$\psi_i = \sum_{j=1}^{n} c_{ij} \, |p_j\rangle \tag{1}$$

where $|p_j\rangle$ symbolizes a p_π-orbital located on the j-th atom of the conjugated molecule, and the summation goes over all n atoms which participate in the conjugation.

The SCHRÖDINGER equation in the HMO theory reads

$$\mathscr{H}\psi_i = E_i\psi_i \tag{2}$$

$i = 1, 2, \ldots, n$, where E_i is the energy of a π-electron in the i-th molecular orbital and where the Hamiltonian operator $\mathscr{H}$ is defined by means of the matrix elements

$$\langle p_j |\mathscr{H}| p_j\rangle = \alpha_j \tag{3}$$

$$\langle p_j |\mathscr{H}| p_k\rangle = \beta_{jk}, \qquad j \neq k. \tag{4}$$

Hence, $\mathscr{H}$ is an effective one electron operator. In the standard HÜCKEL theory, the following apparently drastic approximations are accepted:

$$\alpha_j = \alpha \tag{5}$$

for all atoms $j = 1, 2, \ldots, n$,

$$\beta_{jk} = \beta \tag{6}$$

for all pairs of atoms j, k between which a chemical bond exists, and

$$\beta_{jk} = 0 \tag{7}$$

for all pairs of atoms j, k which are chemically not bonded.

The above relations are assumed to hold only for conjugated *hydrocarbons*. In the case of conjugated molecules containing heteroatoms, not all α-integrals are supposed to be equal and sometimes the non-zero β-integrals are also taken to be different for different sorts of atoms involved in the corresponding chemical bond. These details will not be considered here any further and we shall continue with the outline of the HÜCKEL approximation for hydrocarbons.

In addition to the above requirements for the integrals α and β, the basis orbitals $|p_j\rangle$ are considered as normalized:

$$\langle p_j|p_j\rangle = 1 \tag{8}$$

for all $j = 1, 2, \ldots, n$ (a commonly accepted property of orbitals) and orthogonal:

$$\langle p_j|p_k\rangle = 0 \tag{9}$$

whenever $j \neq k$.

It is very difficult, if not impossible, to give an acceptable physical justification for the approximations (5), (6), (7) and (9). In the early thirties, when HMO theory was invented, the drastic simplifications (5), (6), (7) and (9) were inevitable bacause of the lack of computing machines. In the meantime this excuse for the usage of (5), (6), (7) and (9) can no longer be put forward. Nevertheless, HMO theory still continues to live in theoretical chemistry, although much overshadowed by more sophisticated quantum-chemical methods. As we shall see later on, the role of HMO theory in the modern theoretical chemistry is restricted to the finding of qualitative and, in the best case, semiquantitative predictions of the behaviour of conjugated π-electron systems. However, if the results obtained from the HMO model are used and interpreted in a reasonable manner, then we are again and again surprised how such a simple approach, based on so crude approximations is in a tolerable agreement with experimental findings.

The reasons for this unexpected success of the HMO model are not completely clear. One of them is certainly the close relation between the HMO Hamiltonian and molecular topology.

The connection between the HMO Hamiltonian and the adjacency matrix of a certain graph is the basis of an extensive application of graph (spectral) theory in chemistry and resulted in more than one thousand publications. A plethora of books and reviews covers and elaborates this matter [3, 6, 16, 18, 24, 30, 31, 65] and therefore in this chapter we shall be as concise as possible.

Consider, as an example, pentalene

and write down its HMO Hamiltonian matrix. Applying (5)–(7) we obtain

$$H(\text{pentalene}) = \begin{pmatrix} \alpha & \beta & 0 & 0 & 0 & 0 & \beta & 0 \\ \beta & \alpha & \beta & 0 & 0 & 0 & 0 & 0 \\ 0 & \beta & \alpha & 0 & 0 & 0 & 0 & \beta \\ 0 & 0 & 0 & \alpha & \beta & 0 & 0 & \beta \\ 0 & 0 & 0 & \beta & \alpha & \beta & 0 & 0 \\ 0 & 0 & 0 & 0 & \beta & \alpha & \beta & 0 \\ \beta & 0 & 0 & 0 & 0 & \beta & \alpha & \beta \\ 0 & 0 & \beta & \beta & 0 & 0 & \beta & \alpha \end{pmatrix}. \tag{10}$$

The matrix (10) is the representation of the operator $\mathcal{H}$ for pentalene in the basis formed by the p_π-atomic orbitals.

Using elementary matrix-theoretical transformations (see Appendix 1), we can write $H(\text{pentalene})$ in the form:

$$H(\text{pentalene}) = \alpha \begin{pmatrix} 1 & 0 & 0 & 0 & 0 & 0 & 0 & 0 \\ 0 & 1 & 0 & 0 & 0 & 0 & 0 & 0 \\ 0 & 0 & 1 & 0 & 0 & 0 & 0 & 0 \\ 0 & 0 & 0 & 1 & 0 & 0 & 0 & 0 \\ 0 & 0 & 0 & 0 & 1 & 0 & 0 & 0 \\ 0 & 0 & 0 & 0 & 0 & 1 & 0 & 0 \\ 0 & 0 & 0 & 0 & 0 & 0 & 1 & 0 \\ 0 & 0 & 0 & 0 & 0 & 0 & 0 & 1 \end{pmatrix} + \beta \begin{pmatrix} 0 & 1 & 0 & 0 & 0 & 0 & 1 & 0 \\ 1 & 0 & 1 & 0 & 0 & 0 & 0 & 0 \\ 0 & 1 & 0 & 0 & 0 & 0 & 0 & 1 \\ 0 & 0 & 0 & 0 & 1 & 0 & 0 & 1 \\ 0 & 0 & 0 & 1 & 0 & 1 & 0 & 0 \\ 0 & 0 & 0 & 0 & 1 & 0 & 1 & 0 \\ 1 & 0 & 0 & 0 & 0 & 1 & 0 & 1 \\ 0 & 0 & 1 & 1 & 0 & 0 & 1 & 0 \end{pmatrix}. \tag{11}$$

The first matrix on the right-hand side of the above equation is the unit matrix of order eight. The second matrix is another symmetric matrix of order eight, whose diagonal is zero and whose off-diagonal elements are zero or unity. Bearing in mind what we already know about graphs, we can interpret the matrix in question as the adjacency matrix of a certain graph. This graph is easily recognized as the HÜCKEL graph of pentalene:

Hence we showed that

$$H(\text{pentalene}) = \alpha I + \beta A(\text{HÜCKEL graph of pentalene}). \tag{12}$$

The result obtained is, of course, quite general. If M is a conjugated hydrocarbon and G is its HÜCKEL graph, then

$$H(M) = \alpha I + \beta A(G). \tag{13}$$

The identity (13) is the mathematical expression of the relation which exists between HMO theory and graph theory. Equation (13) has numerous remarkable consequences.

Firstly, the matrices H and A commute, i.e. $HA = AH$. Therefore H and A have identical eigenvectors. In other words, the LCAO coefficients c_{ij} given in Eq. (1) coincide with the components of the i-th eigenvector of the pertinent HÜCKEL graph.

Secondly, if λ_i is an eigenvalue of A, then $\alpha + \beta\lambda_i$ is an eigenvalue of H. Bearing in mind Eq. (2), we see that the eigenvalues of H are the molecular orbital energy levels, i.e.

$$E_i = \alpha + \beta\lambda_i \tag{14}$$

for $i = 1, 2, \dots, n$. This means that the HÜCKEL molecular orbital energies are linear functions of graph eigenvalues.

These two corollaries of Eq. (13) show how important a role the graph eigenvectors and eigenvalues are playing in HMO theory. Sometimes is even claimed that the HÜCKEL theory is fully equivalent to the graph spectral theory (see p. 65 of [65]). This is an exaggeration: graph spectral theory provides only the mathematical basis of the HÜCKEL model.

Several mathematical details related to HMO theory will be elaborated in the next chapter. In Chaps. 12 and 13 two special topics in HMO theory will be presented in full detail. In Chap. 13 also another molecular orbital model will be outlined, in which σ MO energy levels are related to the eigenvalue of the line graph of the molecular graph.

Chapter 6

Special Molecular Graphs

6.1 Acyclic Molecules

In graph theory a connected acyclic graph is called a tree. Hence we may say that the topology of acyclic molecules is represented by trees; the essential topological properties of acyclic molecules coincide with those of trees. In the following we shall get acquainted with the basic properties of trees.

6.1.1 Trees

It is a common joke among graph-theoreticians that if a certain problem cannot be solved for graphs, one should try to solve it for trees. As a matter of fact, trees can be regarded as having the simplest structure among all graphs. On Fig. 6.1 are presented all the fourteen trees with not more than six vertices.

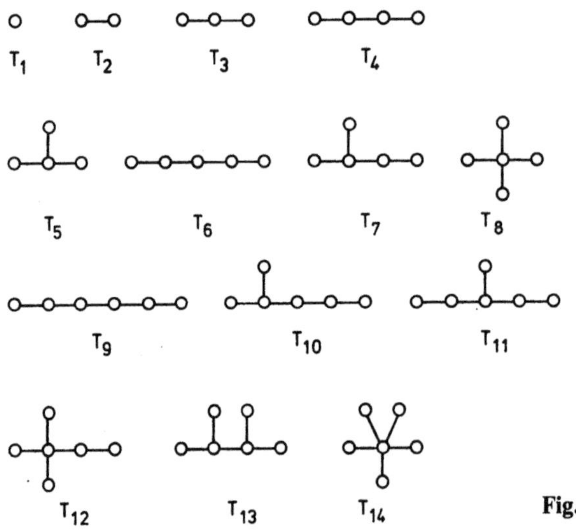

Fig. 6.1. Trees with six or less vertices

Note that the tree with one, two and three vertices is unique. For $n \geq 4$ there are several trees with n vertices and their number rapidly increases with n. We shall not be concerned with the important (solved) problem of the enumeration of trees which is closely related to the famous (and also solved) problem of the enumeration of isomeric alkanes.

The following theorem characterizes the trees.

Theorem 6.1. *Let G be a graph with n vertices and m edges. Then the statements (a)–(e) are equivalent.*
(a) G is a tree, i.e. G is connected and acyclic.
(b) G is acyclic and $m = n - 1$.
(c) G is connected and $m = n - 1$.
(d) G is connected, but $G - e$ is not connected, where e is an arbitrary edge of G.
(e) G is acyclic and every graph obtained by introducing a new edge in G is cyclic.

Note that the statement (d) cannot be applied to T_1 while the statement (e) has two exceptions: T_1 and T_2.

Whenever possible we will denote a tree by T. The set of all trees is denoted by $\mathcal{T}$ whereas the set of all trees with n vertices by $\mathcal{T}_n$. Thus, for instance, $\mathcal{T}_3 = \{T_3\}$ and $\mathcal{T}_6 = \{T_9, T_{10}, T_{11}, T_{12}, T_{13}, T_{14}\}$ (see Fig. 6.1).

As an exercise we prove an elementary result. A vertex of degree one is called a *pendent vertex*.

Lemma 6.2. *Every tree with at least two vertices has at least two pendent vertices.*

Proof. A tree T with more than two vertices has at least one edge. Let e be an edge of T and let its end vertices be u and v.

We prove first that T has a pendent vertex. Suppose that neither u nor v are of degree one and consider the vertex u. It must have at least another neighbour, say u_1. The vertex u_1 is either pendent or has another neighbour, say u_2. The vertex u_2 differs from v since otherwise T would possess a cycle. The vertex u_2 is either pendent or has another neighbour, say u_3. The vertex u_3 differs from v and u_1 since otherwise T would possess a cycle.

Continuing this reasoning and having in mind that the number of vertices of T is finite, we will necessarily come to a vertex which is of degree one.

Applying the same argument to the vertex v we can prove the existence of another pendent vertex in T. $\square$

We have given the proof of Lemma 6.1 in full length because it implies a further, rather important conclusion.

Theorem 6.3. *Every two vertices in a tree are connected by exactly one elementary path.*

For the definition of an elementary path see the paragraph 4.1.4.

6.1.2 The Path and the Star

With regard to Lemma 6.2 we may be interested to find trees with minimal ($= 2$) and maximal number of pendent vertices. The answer is simple: in $\mathcal{T}_n$ there is a unique tree with two pendent vertices (called the path and denoted by P_n) and another unique tree with $n - 1$ pendent vertices (called the star and denoted by $K_{1,n-1}$). Their structure and the way in which we shall label their vertices is given as follows:

$$P_n \qquad\qquad K_{1,n-1}$$

Among the trees from Fig. 6.1, T_1, T_2, T_3, T_4, T_6 and T_9 are the paths with $n = 1, 2, 3, 4, 5$ and 6 vertices. Furthermore, $T_2 \simeq K_{1,1}$, $T_3 \simeq K_{1,2}$, $T_5 \simeq K_{1,3}$, $T_8 \simeq K_{1,4}$ and $T_{14} \simeq K_{1,5}$.

6.1.3 The Characteristic Polynomial of Trees

Applying the SACHS theorem (see the paragraph 4.3.3) to trees we easily obtain the following conclusion. Since trees do not contain cycles, $c(S) = 0$ must hold for all their SACHS graphs or, in other words, every SACHS graph of a tree is composed of components K_2, exclusively. As already pointed out (see the paragraph 4.2.2), a subgraph, composed exclusively of components K_2 represents a matching. We see, therefore, that in the case of trees there is a one-to-one correspondence between a SACHS graph with $2k$ vertices and a k-matching.

The number of k-matchings of a graph G is denoted by $m(G, k)$.

Theorem 6.4a. *For all $T \in \mathcal{T}$ the coefficients of the characteristic polynomial of T obey the relations*

$$a_{2k}(T) = (-1)^k\, m(T, k) \tag{1}$$

and

$$a_{2k+1}(T) = 0 \tag{2}$$

for all $k \geqq 0$.

Using the definition of the matching polynomial [see Eq. (4.11)] we arrive at another form of the above theorem.

Theorem 6.4b. *For all $T \in \mathcal{T}$ the characteristic and the matching polynomial coincide:*

$$\varphi(T) = \alpha(T) \tag{3}$$

i.e.

$$\varphi(T, x) = \sum_{k=0}^{[n/2]} (-1)^k\, m(T, k)\, x^{n-2k}. \tag{4}$$

It can be shown that Eq. (4) holds only for acyclic graphs. Namely if a graph G contains cycles, then the statement of Theorem 6.4a is violated for at least one value of k and consequently, the characteristic and the matching polynomial of G differ.

Using the basic recurrence for the number of k-matchings [Eqs. (4.9) and (4.10)], we immediately attain a few recurrence relations for the characteristic polynomial of a tree.

Theorem 6.5. *If e is any edge of $T \in \mathcal{T}$, connecting the vertices u and v, then*

$$\varphi(T) = \varphi(T - e) - \varphi(T - u - v) \,. \tag{5}$$

Note that because of part (d) of Theorem 6.1 the subgraph $T - e$ is always composed of two components, which are trees themselves. The number of components of $T - u - v$ is equal to $g_u + g_v - 2$, where g denotes the vertex degree. Supposing that $T - e = T_a \cup T_b$, we can further simplify Eq. (5), viz.

$$\varphi(T) = \varphi(T_a)\, \varphi(T_b) - \varphi(T_a - u)\, \varphi(T_b - v) \,. \tag{6}$$

There is a straightforward generalization of Theorem 6.5. An edge of a graph G is called a *bridge* if $G - e$ possesses more components than G. [By part (d) of Theorem 6.1, all edges of trees are bridges.] Then if e is a bridge of $G \in \mathcal{G}$, connecting the vertices u and v,

$$\varphi(G) = \varphi(G - e) - \varphi(G - u - v) \,. \tag{7}$$

If, in addition, $G - e = G_a \cup G_b$, then

$$\varphi(G) = \varphi(G_a)\, \varphi(G_b) - \varphi(G_a - u)\, \varphi(G_b - v) \,. \tag{8}$$

An important special case of formula (8) is obtained when G_a (or G_b) is a graph possessing just one vertex. Then $\varphi(G_a) = x$ and $\varphi(G_a - u) = 1$.

Corollary 6.5.1. *If v is a pendent vertex of $G \in \mathcal{G}$, being adjacent to the vertex u, then*

$$\varphi(G) = x\varphi(G - v) - \varphi(G - v - u) \,. \tag{9}$$

Recall that by Lemma 6.2 pendent vertices necessarily exist in trees.

Formulas (5)–(9) enable a very easy and efficient calculation of the characteristic polynomial of trees and, more generally, of graphs having bridges and pendent vertices [74, 136].

Corollary 6.5.2. *If v is a vertex of $T \in \mathcal{T}$, being adjacent to the vertices $u_1, u_2, \ldots, u_g$, then*

$$\varphi(T) = x\varphi(T - v) - \sum_{i=1}^{g} \varphi(T - u_i - v) \,. \tag{10}$$

Proof. Apply (5) repeatedly to the edges $e_1, e_2, \ldots, e_g$ connecting v with $u_1, u_2, \ldots, u_g$ and note that $T - e_1 - e_2 - \ldots - e_g \simeq K_1 \cup T - v$, where K_1 is the graph possessing just one vertex. Then

$$\varphi(T - e_1 - e_2 - \ldots - e_g) = \varphi(K_1)\,\varphi(T - v) = x\varphi(T - v)\,. \qquad \Box \tag{11}$$

In order to illustrate the above recurrence formulas, we shall determine the characteristic polynomial of T^*:

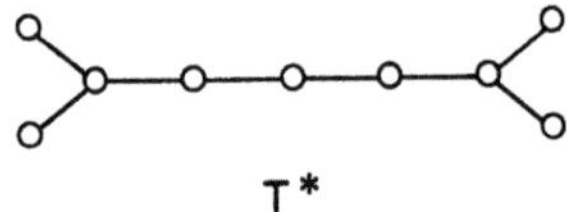

$$T^*$$

Application of Theorem 6.5 on the edge indicated by an arrow gives

$$\phi\left(\text{graph}\right) = \phi\left(\text{graph}\right) - \phi\left(\text{graph}\right) =$$

$$= \phi\left(\text{graph}\right) \cdot \phi\left(\text{graph}\right) - \phi\left(\text{graph}\right) \cdot \phi\left(\text{graph}\right) =$$

$$= \phi(T_5)\,[\phi(T_7) - \phi(T_3)] \tag{12}$$

where the symbols T_3, T_5 and T_7 refer to Fig. 6.1. It is now easy to calculate that

$$\varphi(T_3) = x^3 - 2x \tag{13}$$

$$\varphi(T_5) = x^4 - 3x^2 \tag{14}$$

$$\varphi(T_7) = x^5 - 4x^3 + 2x\,. \tag{15}$$

For instance, using Eq. (9) we have

$$\phi\left(\text{graph}\right) = x\,\phi\left(\text{graph}\right) - \phi\left(\text{graph}\right) =$$

$$= x(x^4 - 3x^2) - (x^3 - 2x) \tag{16}$$

Substituting these polynomials back into the above expression for $\varphi(T^*)$ we finally get

$$\varphi(T^*) = (x^4 - 3x^2)\,[(x^5 - 4x^3 + 2x) - (x^3 - 2x)] = x^9 - 8x^7 + 19x^5 - 12x^3 \tag{17}$$

or in factorized form

$$\varphi(T^*) = x^3(x^2 - 1)\,(x^2 - 3)\,(x^2 - 4)\,. \tag{18}$$

6.1.4 Trees with Greatest Number of Matchings

In certain applications, which are discussed in Chaps. 11 and 12, the comparison of trees by their matching numbers will be needed. Let us therefore pose the question, which trees (with a given number of vertices) have the maximum and minimum number of matchings.

Since not all matchings are of the same cardinality, it is useful to introduce a quasiordering of graphs as follows [100, 104, 113]. Let $m(G, k)$ be the number of k-matchings of the graph G. If for two graphs G_1 and G_2 the inequalities

$$m(G_1, k) \geqq m(G_2, k) \tag{19}$$

hold for all values of k, then we write $G_1 \succ G_2$ or $G_2 \prec G_1$. If both $G_1 \succ G_2$ and $G_2 \succ G_1$ we write $G_1 \sim G_2$ and say that the graphs G_1 and G_2 are matching equivalent. Note that for the definition of the quasiordering $G_1 \succ G_2$ it is not necessary that the graphs G_1 and G_2 have an equal number of vertices.

If neither $G_1 \succ G_2$ nor $G_1 \prec G_2$ hold then the graphs G_1 and G_2 are said to be matching incomparable and we shall write $G_1 \nsim G_2$. There exist matching equivalent as well as matching incomparable pairs of trees, as shown by the following examples.

$$m(T, 1) = 7 \qquad\qquad m(T, 1) = 7$$
$$m(T, 2) = 9 \qquad\qquad m(T, 2) = 9$$
$$m(T, 3) = 0 \qquad\qquad m(T, 3) = 0$$

$$m(T, 1) = 7 \qquad\qquad m(T, 1) = 7$$
$$m(T, 2) = 14 \qquad\qquad m(T, 2) = 12$$
$$m(T, 3) = 8 \qquad\qquad m(T, 3) = 7$$
$$m(T, 4) = 0 \qquad\qquad m(T, 4) = 1$$

Nevertheless, in many cases trees are matching comparable.

Theorem 6.6. *If* $T \in \mathscr{T}_n$, *then*

$$K_{1, n-1} \prec T \prec P_n . \tag{20}$$

Furthermore, T is matching equivalent to $K_{1, n-1}$ and P_n only if $T \simeq K_{1, n-1}$ and $T \simeq P_n$, respectively.

Proof. For all trees $T \in \mathscr{T}_n$, $m(T, 1) = n - 1$. Now it is obvious that in the star no two edges are mutually independent. Hence $m(K_{1, n-1}, k) = 0$ for all $k \geq 2$. For all other trees it is possible to find at least one pair of independent edges. Hence $K_{1,n-1} \prec T$.

We prove the second part of the theorem by induction on the number of vertices. Since $\mathcal{T}_2 = \{T_2\}$ and $\mathcal{T}_3 = \{T_3\}$, the statement $T \prec P_n$ is in a trivial manner fulfilled for $n = 2$ and $n = 3$.

Suppose now that the statement is true for all trees with $n - 1$ and $n - 2$ vertices. Consider an arbitrary tree $T \in \mathcal{T}_n$. Let v be a pendent vertex of T. Lemma 6.2 guarantees that such a vertex exists. Let u be the vertex adjacent to v. Then by (4.10),

$$m(T, k) = m(T - v, k) + m(T - u - v, k - 1). \tag{21}$$

A special case of this identity is

$$m(P_n, k) = m(P_{n-1}, k) + m(P_{n-2}, k - 1). \tag{22}$$

It is clear that $T - v \in \mathcal{T}_{n-1}$ and $T - u - v \in \mathcal{T}_{n-2}$. Then according to the induction hypothesis

$$m(T - v, k) \leqq m(P_{n-1}, k) \tag{23}$$

and

$$m(T - u - v; k - 1) \leqq m(P_{n-2}, k - 1) \tag{24}$$

for all k. Summing these two inequalities we get

$$m(T - v, k) + m(T - u - v, k - 1) \leqq m(P_{n-1}, k) + m(P_{n-2}, k - 1). \tag{25}$$

This implies,

$$m(T, k) \leqq m(P_n, k) \tag{26}$$

and thus we have also proved the right-hand part of Theorem 4.6.   □

It has been shown that in the set $\mathcal{T}_n \setminus \{K_{1,n-1}, P_n\}$ the trees with minimum and maximum number of matchings are T_a and T_d, respectively [100].

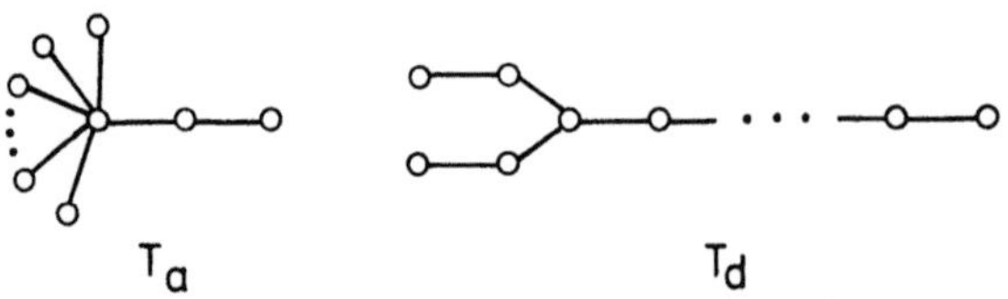

Monocyclic [104], bicyclic [104] and tricyclic [113] graphs with maximum number of matchings have also been determined.

We now formulate an auxiliary result which will be needed later on.

Lemma 6.7. *Let $n = 4k$ or $4k + 1$ or $4k + 2$ or $4k + 3$. Then*

$$P_n \succ P_2 \cup P_{n-2} \succ P_4 \cup P_{n-4} \succ \ldots \succ P_{2k} \cup P_{n-2k} \succ P_{2k+1} \cup P_{n-2k-1}$$
$$\succ P_{2k-1} \cup P_{n-2k+1} \succ \ldots \succ P_3 \cup P_{n-3} \succ P_1 \cup P_{n-1} . \tag{27}$$

Proof can also be performed by induction on the number of vertices of $P_j \cup P_{n-j}$. Since the procedure is quite lengthy and not too interesting, we present only a characteristic detail of it.

Suppose that we want to show that for $n \geq 2$, $P_2 \cup P_{n-2} \succ P_1 \cup P_{n-1}$. For $n = 2$ and $n = 3$ we can check this statement by direct calculation. Assuming the validity of $P_2 \cup P_{n-3} \succ P_1 \cup P_{n-2}$ and $P_2 \cup P_{n-4} \succ P_1 \cup P_{n-3}$, i.e.

$$m(P_2 \cup P_{n-3}, k) \geq m(P_1 \cup P_{n-2}, k) \tag{28}$$

and

$$m(P_2 \cup P_{n-4}, k - 1) \geq m(P_1 \cup P_{n-3}, k - 1) \tag{29}$$

we obtain

$$m(P_2 \cup P_{n-3}, k) + m(P_2 \cup P_{n-4}, k - 1) \geq m(P_1 \cup P_{n-2}, k)$$
$$+ m(P_1 \cup P_{n-3}, k - 1) \tag{30}$$

and therefore

$$m(P_2 \cup P_{n-2}, k) \geq m(P_1 \cup P_{n-1}, k) . \quad \square \tag{31}$$

6.1.5 The Spectrum of the Path

In numerous chemical applications the spectrum and/or the characteristic polynomial of the path play an important role. These results have long been known [15, 140]. We shall repeat them here in order to provide a more advanced illustration of the material outlined in the previous sections.

Applying Corollary 6.5.1 we get the simple recursion relation

$$\varphi(P_n) = x\varphi(P_{n-1}) - \varphi(P_{n-2}) \tag{32}$$

which together with the initial conditions $\varphi(P_0) = 1$, $\varphi(P_1) = x$ enables an easy step-by-step calculation of $\varphi(P_n)$.

Solving the relation (32) by standard algebraic methods, we get

$$\varphi(P_n, x) = (x^2 - 4)^{-1/2} \left\{ \left[\frac{1}{2}(x + \sqrt{x^2 - 4}) \right]^{n+1} - \left[\frac{1}{2}(x - \sqrt{x^2 - 4}) \right]^{n+1} \right\} . \tag{33}$$

Setting $x = 2 \cos \tau$ and performing appropriate trigonometric transformations, this formula is reduced to

$$\varphi(P_n, x) = \sin (n + 1) \, \tau / \sin \tau \,. \tag{34}$$

The zeros of $\varphi(P_n, x)$ can now be easily deduced. They read

$$\lambda_j = 2 \cos [j\pi/(n + 1)] \tag{35}$$

$j = 1, 2, \ldots, n$. These are, of course, also the eigenvalues of P_n. The eigenvector of P_n, corresponding to the eigenvalue λ_j has components

$$c_{jr} = \sqrt{2(n + 1)} \, \sin [jr\pi/(n + 1)] \tag{36}$$

$r = 1, 2, \ldots, n$.

It can be further shown that

$$m(P_n, k) = \binom{n - k}{k} \tag{37}$$

and consequently

$$\varphi(P_n, x) = \sum_{k=0}^{[n/2]} (-1)^k \binom{n - k}{k} x^{n - 2k} \,. \tag{38}$$

Since the CHEBYSHEV polynomial of the second kind is defined as

$$U_n(x) = (1 - x^2)^{-1/2} \sin [(n + 1) \arccos x] \tag{39}$$

the characteristic polynomial of the path obeys

$$\varphi(P_n, x) = U_n(x/2) \,. \tag{40}$$

6.2 The Cycle

As already mentioned in paragraph 4.1.6 the cycle is a connected regular graph of degree two. In the same section the first five cycles are given as examples. The cycle with n vertices is denoted by C_n.

The cycle has the following obvious properties. If e is any edge of C_n, then $C_n - e \simeq P_n$. If v is any vertex of C_n, then $C_n - v \simeq P_{n-1}$. Furthermore, if u and v are adjacent vertices of C_n, then $C_n - u - v \simeq P_{n-2}$.

Bearing this in mind, we can write down the following special case of Eq. (4.31)

$$\varphi(C_n) = \varphi(P_n) - \varphi(P_{n-2}) - 2 \,. \tag{41}$$

Of course, $G - Z$ in formula (4.31) is now the graph without vertices and its characteristic polynomial is by definition equal to unity.

Substituting (34) into (41) one arrives at

$$\varphi(C_n, x) = 2 \cos n\tau - 2 \tag{42}$$

where $x = 2 \cos \tau$.

The eigenvalues of C_n are now obtained by equating $\varphi(C_n)$ to zero. Since $\cos \alpha = 1$ for $\alpha = 2j\pi$, from (42) we conclude that $\varphi(C_n, x) = 0$ whenever $\tau = 2j\pi/n$ and hence $x = 2 \cos (2j\pi/n)$. Consequently,

$$x_j = 2 \cos (2j\pi/n) \tag{43}$$

$j = 1, 2, \ldots, n$ are the eigenvalues of C_n.

It can be demonstrated that

$$\varphi(C_n, x) = \sum_{k=0}^{[n/2]} (-1)^k \frac{n}{n-k} \binom{n-k}{k} x^{n-2k} - 2 \,. \tag{44}$$

The CHEBYSHEV polynomial of the first kind is defined as

$$T_n(x) = \cos [n \arccos x] \,. \tag{45}$$

Combining (42) with (45) we immediately see that

$$\varphi(C_n, x) = 2T_n(x/2) - 2 \,. \tag{46}$$

6.3 Alternant Molecules

The concept of the alternant and non-alternant hydrocarbons was introduced by COULSON and RUSHBROOKE [80] in connection with the discovery of the pairing theorem. Over decades this was the most important result in molecular orbital theory which had a clearly topological origin.

COULSON and RUSHBROOKE noticed, namely, that it is reasonable to divide conjugated hydrocarbons into two classes, depending upon whether or not their carbon atoms can be labeled by stars and circles, so that first neighbours carry different labels. Such a classification of unsaturated cyclic hydrocarbons eventually became standard in organic chemistry.

For example, naphthalene and biphenylene are alternant hydrocarbons as one can check from the following "starred" diagrams:

On the other hand, heptafulvene and dicalicene are non-alternant since the atoms marked by a question mark cannot be labeled by either a star or a circle:

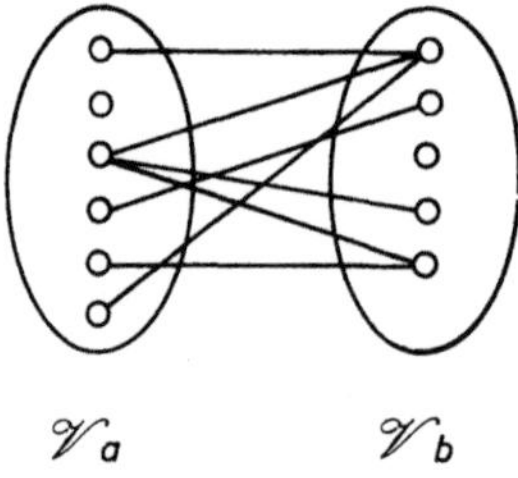

The reason why such a classification is important will be seen from what follows.

6.3.1 Bipartite Graphs

In graph theory the notion of a bipartite graph precisely corresponds to the chemical concept of an alternant hydrocarbon. A *bipartite graph* is a graph whose vertex set can be partitioned into two parts, say $\mathscr{V}_a$ and $\mathscr{V}_b$, such that two adjacent vertices belong neither both to $\mathscr{V}_a$ nor both to $\mathscr{V}_b$. Hence edges exist only between vertices belonging to different sets:

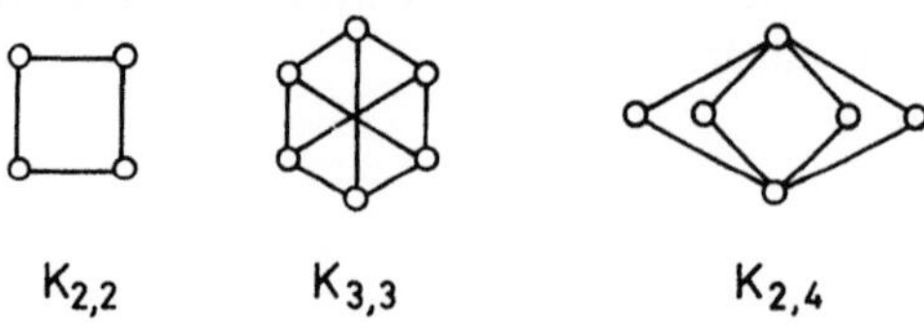

Clearly, if one labels the vertices of $\mathscr{V}_a$ by stars and the vertices of $\mathscr{V}_b$ by circles, then two adjacent vertices will always carry different labels.

The following characterization of bipartite graphs is important.

Theorem 6.8. *A graph is bipartite if and only if it contains no odd-membered cycles.*

We shall adopt the following conventions. The sets $\mathscr{V}_a$ and $\mathscr{V}_b$ have a and b elements, respectively and $a \leq b$. The bipartite graph whose vertex set is $\mathscr{V}_a \cup \mathscr{V}_b$ will be said to have $a + b$ vertices and to belong to the set $\mathscr{G}_{a,b}$. The element of $\mathscr{G}_{a,b}$ having the maximum possible number of edges is called the complete bipartite graph on $a + b$ vertices and is denoted by $K_{a,b}$.

It should be an easy exercise for the reader to convince himself that the following graphs are $K_{2,2}$, $K_{3,3}$ and $K_{2,4}$.

$$K_{2,2} \qquad K_{3,3} \qquad K_{2,4}$$

The star is a further example of a complete bipartite graph.

6.3.2 The Pairing Theorem

Theorem 6.9 *(the Pairing Theorem). If $\lambda_1, \lambda_2, \ldots, \lambda_n$ are the eigenvalues of a bipartite graph, then*

$$\lambda_i = -\lambda_{n-i+1} \tag{47}$$

for all $i = 1, 2, \ldots, n$. In other words, the eigenvalues of a bipartite graph occur in pairs of the form $(\lambda, -\lambda)$. Exceptionally, if n is odd, one eigenvalue remains unpaired; its value is equal to zero.

First proof. According to Theorem 6.8 a bipartite graph G does not possess odd-membered cycles. Then, of course, no SACHS graph of G possesses odd cycles. Hence all SACHS graphs of G are composed of graphs K_2 and/or even-membered cycles. Hence all SACHS graphs of G have an even number of vertices.

From the SACHS theorem (see paragraph 4.3.3) is now evident that for a bipartite graph all odd coefficients of the characteristic polynomial are equal to zero and therefore the characteristic polynomial takes the form

$$\varphi(G, x) = \sum_{k=0}^{[n/2]} a_{2k} x^{n-2k} . \tag{48}$$

Consequently, $\varphi(G, -x) = (-1)^n \varphi(G, x)$ and if for some λ, $\varphi(G, \lambda) = 0$, then also $\varphi(G, -\lambda) = 0$.

This proves the theorem. $\square$

Second proof. Label the vertices of G so that the vertices $1, 2, \ldots, a$ belong to $\mathscr{V}_a$ and the vertices $a + 1, a + 2, \ldots, a + b = n$ belong to $\mathscr{V}_b$. Then the adjacency matrix of G has the following block form:

$$A = \begin{pmatrix} 0_a & B \\ B^\dagger & 0_b \end{pmatrix} \tag{49}$$

where B is an $a \times b$ matrix, $B^\dagger$ is its transpose and 0_a and 0_b are zero-matrices of order a and b, respectively.

We can now write the eigenvalue-eigenvector equation as

$$\begin{pmatrix} 0_a & B \\ B^\dagger & 0_b \end{pmatrix} \begin{pmatrix} C_a \\ C_b \end{pmatrix} = \lambda \begin{pmatrix} C_a \\ C_b \end{pmatrix} \tag{50}$$

where we have omitted the index i from λ_i and C_i. Besides, the eigenvector C is presented in a block form

$$C = \begin{pmatrix} C_a \\ C_b \end{pmatrix} \tag{51}$$

where C_a and C_b are vectors of dimension a and b, respectively.

After performing the matrix multiplications in (50) we obtain two equations

$$BC_b = \lambda C_a \tag{52}$$

and

$$B^\dagger C_a = \lambda C_b \tag{53}$$

which must be simultaneously obeyed. Now, these equations are not changed if λ is replaced by $-\lambda$ and C_b by $-C_b$. Hence, if λ is an eigenvalue of G, then also $-\lambda$ is an eigenvalue. $\square$

We have, however, proved more.

Corollary 6.9.1. *If $[C_a, C_b]^\dagger$ is the eigenvector of $G \in \mathcal{G}_{a,b}$, corresponding to the eigenvalue λ, then $[C_a, -C_b]^\dagger$ is the eigenvector corresponding to the eigenvalue $-\lambda$. Hence also the eigenvectors of bipartite graphs are paired.*

6.3.3 Some Consequences of the Pairing Theorem

In order to simplify the analysis which follows, consider a bipartite graph $G \in \mathcal{G}_{a,a}$. Then according to Corollary 6.9.1 the unitary matrix which diagonalizes the adjacency matrix of G (see Appendix 4) has the following block form

$$U = \begin{pmatrix} X & X \\ Y & -Y \end{pmatrix} \tag{54}$$

where X and Y are square matrices or order a. The condition $U^\dagger U = I_{2a}$ implies

$$2X^\dagger X = 2Y^\dagger Y = I_a . \tag{55}$$

Within the HÜCKEL molecular orbital theory the diagonal elements of the matrices $2X^\dagger X$ and $2Y^\dagger Y$ are interpreted as the π-electron charge densities on the atoms labeled by stars and circles, respectively, of an alternant conjugated hydrocarbon [123]. From the above relations we see that the π-electron charge density on all atoms is equal to unity. This is just the well-known result of the HÜCKEL theory that all alternant hydrocarbons have a uniform π-electron charge distribution [80].

A detailed analysis would show that the same conclusion is reached for a graph $G \in \mathcal{G}_{a,b}$, if $a \neq b$.

Equation (48), deduced in the first proof of the pairing theorem, has a further consequence.

Corollary 6.9.2. *If $G \in \mathcal{G}_{a,b}$, then the characteristic polynomial of G can be written in the form*

$$\varphi(G, x) = \sum_{k=0}^{a} (-1)^k \, b(G, k) \, x^{a+b-2k} , \tag{56}$$

where $b(G, k) \geq 0$ for all values of k.

Proof. Suppose in the spectrum of G there are n_+ positive, n_0 zero and n_- negative eigenvalues. Because of the pairing theorem, $n_+ = n_-$. Therefore, only the coefficients $a_0 = 1, a_2, a_4, \ldots, a_{2n_+}$ in (48) are different from zero.

According to the DESCARTES theorem (see Appendix 4), there must be exactly n_+ sign changes in the sequence $a_0, a_2, a_4, \ldots, a_n$. Obviously this is possible only if $a_0 > 0$, $a_2 < 0$, $a_4 > 0$ etc. Consequently, we can write $a_{2k} = (-1)^k b(G, k)$ where $b(G, k) > 0$ for $k = 0, 1, \ldots, n_+$ and $b(G, k) = 0$ for $k > n_+$.

In order to complete the proof of Corollary 6.9.2, note that because of (49), n_+ is equal to the rank of $\boldsymbol{B}$, and obviously rank $\boldsymbol{B} \leq a$. Therefore, the summation in (56) needs not to go beyond $k = a$. $\square$

6.4 Benzenoid Molecules

Benzenoid hydrocarbons form an important and well investigated class of unsaturated conjugated compounds. The parent compound — benzene — was discovered as early as 1825 and in the meantime almost 500 hydrocarbons containing condensed benzene rings have been obtained.

Figure 6.2 shows some representatives of benzenoid hydrocarbons together with their trivial names.

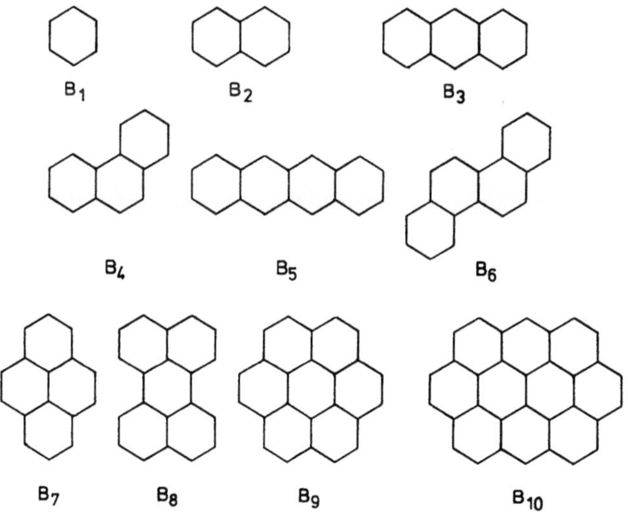

Fig. 6.2. Examples of benzenoid systems. B_1 = benzene, B_2 = naphthalene, B_3 = anthracene, B_4 = phenanthrene, B_5 = tetracene, B_6 = chrysene, B_7 = pyrene, B_8 = perylene, B_9 = coronene, B_{10} = ovalene

6.4.1 Benzenoid Graphs

Benzenoid graphs are the networks obtained by arranging congruent regular hexagons in the plane, so that two hexagons are either disjoint or possess a common edge.

In the chemical and mathematical literature the names *"polyhex"*, *"hexagonal animal"*, *"hexanimal"*, *"hexagonal polyomino"* and *"hexagonal system"* are also used to describe the same class of graphs. It is obvious that benzenoid graphs provide the proper topological representation of benzenoid hydrocarbons. The objects presented on Fig. 6.2 can be viewed as benzenoid graphs.

A more precise definition of a benzenoid graph is the following. Let Z_B be a cycle on the hexagonal (i.e. graphite) lattice. Then the vertices and the edges which lie on Z_B and in the interior of Z_B form a benzenoid graph B. The cycle Z_B is called the perimeter of B. An example is given on Fig. 6.3.

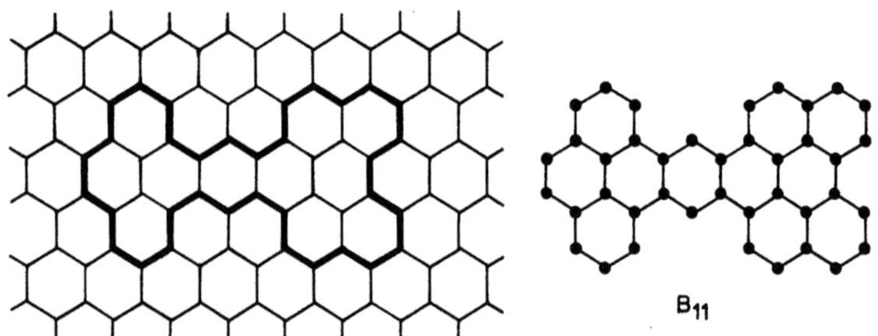

Fig. 6.3. A cycle on the hexagonal lattice and the corresponding benzenoid graph

The vertices of a benzenoid graph can be partitioned into external and internal. The external vertices are those belonging to the perimeter. The rest of the vertices are called internal. The number of internal vertices is denoted by n_i. For the benzenoid graph B_{11} in Fig. 6.3, $n_i = 6$.

Those benzenoid system for which $n_i = 0$ are called *cata-condensed*; those for which $n_i > 0$ are *peri-condensed*. The graphs B_1–B_6 from Fig. 6.2 are examples of cata-condensed benzenoids whereas B_7–B_{10} are peri-condensed.

Let h denote the number of hexagons, n the number of vertices, n_2 and n_3 the number of vertices of degree two and three, and m the number of edges of a benzenoid graph. Then these parameters are related by [97, 160]:

$$n = 4h + 2 - n_i, \tag{57}$$

$$m = 5h + 1 - n_i, \tag{58}$$

$$n_2 = 2h + 4 - n_i, \tag{59}$$

$$n_3 = 2h - 2. \tag{60}$$

HARARY and HARBORTH [125] proved the following result.

Theorem 6.10. *Benzenoid graphs exist whenever the parameters n, m, h and n_i are within the ranges (61)–(66):*

$$2h + 1 + \{\sqrt{12h - 3}\} \leqq n \leqq 4h + 2 \tag{61}$$

$$3h + \{\sqrt{12h - 3}\} \leqq m \leqq 5h + 1 \tag{62}$$

$$n - 1 + \{(n - 2)/4\} \leqq m \leqq 2n - \{[n + \sqrt{6n}]/2\} \tag{63}$$

$$\{(n - 2)/4\} \leqq h \leqq n + 1 - \{[n + \sqrt{6n}]/2\} \tag{64}$$

$$4\{(n - 2)/4\} + 2 - n \leqq n_i \leqq 3n + 6 - 4\{[n + \sqrt{6n}]/2\} \tag{65}$$

$$0 \leqq n_i \leqq 2h + 1 - \{\sqrt{12h - 3}\}\,, \tag{66}$$

where $\{x\}$ denotes the smallest integer being greater than or equal to x. All values of the parameters n, m, h and n_i within the above ranges can occur in benzenoid graphs.

Benzenoid graphs contain no odd-membered cycles. Then by Theorem 6.8 they are bipartite. Benzenoid graphs may possess even-membered cycles of all sizes, except of size 4 and 8. In [81] the following peculiar result has been obtained.

Theorem 6.11. *Let B be a benzenoid graph and Z one of its cycles. If the size of Z is divisible by four, then in the interior of Z there is an odd number of vertices. If the size of Z is not divisible by four, then in the interior of Z there are either no vertices or their number is even.*

Corollary 6.11.1. *All cycles in a cata-condensed benzenoid system have sizes not divisible by four (hence 6, 10, 14, 18 etc.). In every peri-condensed benzenoid system there is at least one cycle, the size of which is divisible by four (hence 12, 16, 20, etc.).*

6.4.2 The Characteristic Polynomial of Benzenoid Graphs

Since a benzenoid graph is bipartite, its characteristic polynomial can be written in the form (56). In the case of benzenoid systems, the coefficients $b(G, k)$ have some special properties. These will be pointed out in the following theorem [84, 107].

Let $m(G, k)$ denote the number of k-matchings of the graph G.

Theorem 6.12. *If G is a benzenoid graph, then*

(a) $b(G, k) = 0$ if and only if $m(G, k) = 0$, $\tag{67}$

(b) $b(G, k) \geqq m(G, k)$ for all $k \geq 0$, $\tag{68}$

and, if n is even, then

(c) $b(G, n/2) = m(G, n/2)^2$. $\tag{69}$

If n is even, then $m(G, n/2)$ is equal to the number of perfect matchings of G, which coincides with the number of KEKULÉ structures of the corresponding conjugated molecule. DEWAR and LONGUET-HIGGINS [84] were the first who observed the important relation

$$\det A = (-1)^{n/2} K^2 \tag{70}$$

which, of course, is equivalent to statement (c) of Theorem 6.12. Here A denotes the adjacency matrix and K the KEKULÉ structure count of the corresponding benzenoid molecule.

Equation (70) has had a great impact on the development of various theoretical approaches to benzenoid hydrocarbons. These have been reviewed elsewhere [28, 29]. Some applications of (70) will be discussed in Chap. 12.

As a consequence of the pairing theorem and Eq. (70) the number of KEKULÉ structures of a benzenoid molecule is obtained by multiplying the non-negative eigenvalues of the molecular graph:

$$K = \prod_{j=1}^{\{n/2\}} \lambda_j . \tag{71}$$

In particular, $K = 0$ if and only if the molecular graph possesses a zero eigenvalue.

6.5 Hydrocarbons and Molecules with Heteroatoms

6.5.1 On the Question of the Molecular Graph

In Chaps. 1 and 2 it was outlined that the molecular graph does not contain information about the nature of the atoms and chemical bonds in the corresponding molecule, that is about the materialization of a given topology. As a consequence of this, quite dissimilar chemical species may fall into the same class of isotopological molecules (cf. Fig. 2.1).

In the case of hydrocarbons this lack of information is fully compensated by our *a priori* knowledge of the type of atoms occuring in such molecules. Furthermore, it is trivially easy to distinguish the vertices representing the carbon atoms from the vertices (if any) representing the hydrogen atoms.

We shall consider as a heteroatom any atom in an organic molecule which is neither carbon nor hydrogen. The type and the position of heteroatoms in a molecule can be deduced only in exceptional cases from its molecular graph.

In order to overcome this shortcoming of the topological description of molecules one needs to supply the molecular graph with some additional information about the atoms and chemical bonds. This can be done in several distinct, but mutually equivalent ways.

The most frequently used graphic representation of molecules with heteroatoms is that in which each heteroatom is symbolized by a vertex with a weighted loop [70, 166]. The weight of the loop is characteristic to the type of the heteroatom.

For instance, the molecular graph of pyridine is given as follows:

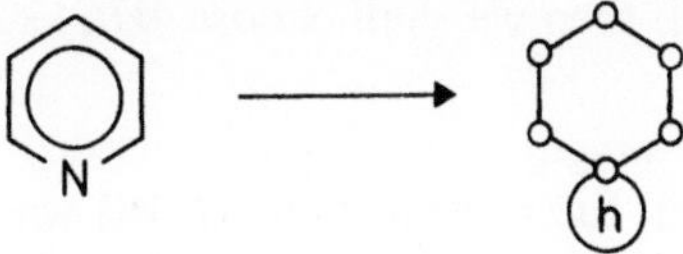

where h is a certain number (= weight of the loop). If we want to indicate that the CN bonds in pyridine are not the same as the CC bonds, we may associate a certain weight, say k, to the edges representing the CN bonds:

Such a representation of molecules with heteroatoms has its obvious origin in the application of graph theory to HÜCKEL molecular orbital theory (see Chap. 5). The weights h and k are, in fact, closely related to the HMO parameters used in the COULOMB and the resonance integrals.

We would like to point out the following disadvantages of such a representation. Firstly, the diagrams called "molecular graphs" are not graphs at all (in the sense of the definitions given in the paragraphs 4.1.2 and 4.1.3). Rather they provide a pictorial representation of a matrix, in particular the HMO Hamiltonian matrix. Secondly, isotopological molecules now have different molecular "graphs". As a consequence of this, the molecular "graph" can be no more viewed as simply depicting the topology of a molecule. Thirdly, the molecular "graph" depends on such marginal facts as the choice of the HMO parameters for the heteroatom.

We propose here a slightly different approach in which the molecules with heteroatoms are also represented by simple graphs [96, 154].

For example, the HÜCKEL graph of pyridine:

is isomorphic to the HÜCKEL graph of benzene. A *weight matrix* $W = W(G)$ is associated with the molecular graph G and by means of this matrix additional information concerning the atoms and the bonds can be recorded. The elements of $W(G)$ are interpreted as weights. In particular if the vertices of G are denoted by $v_1, v_2, \dots, v_n$, then

w_{rr} is the weight of the vertex v_r,
w_{rs} is the weight of the edge, connecting the vertices v_r and v_s.

The weight matrix must be symmetric ($w_{rs} = w_{sr}$) and it is rather convenient to choose $w_{rs} = 0$ whenever the vertices v_r and v_s are not adjacent.

The method of construction of the weight matrix depends exclusively on the specific application for which one will need it. In particular, if we want to use graph theory within the framework of the HMO model, then we shall choose $W(G)$ so as to coincide with the HMO Hamiltonian matrix.

For different purposes a completely different weight matrix may be needed.

If G is a graph and W its weight matrix, then the ordered pair (G, W) will be called a *weighted graph*. The weighted graph is the appropriate topological representation of molecules containing heteroatoms.

6.5.2 The Characteristic Polynomial of Weighted Graphs

Whereas the characteristic polynomial of simple graphs is just the characteristic polynomial of their adjacency matrix, in the case of weighted graphs it is reasonable to consider the characteristic polynomial of the weight matrix. Hence if (G, W) is a weighted graph, then

$$\varphi(G, W) = \varphi(G, W, x) = \det (xI - W) \tag{72}$$

will be its characteristic polynomial.

In the following we shall be interested only in the special case when

$$W = A + \text{diag} (w_1, w_2, \ldots, w_n) \tag{73}$$

where A denotes the adjacency matrix of the graph G and $\text{diag} (w_1, w_2, \ldots, w_n)$ is the diagonal matrix of vertex weights.

Theorem 6.13. *If the relation (73) holds and $w_1 = w_2 = \ldots = w_n = w$, then*

$$\varphi(G, W, x) = \varphi(G, x - w) \, . \tag{74}$$

The verification of the above result should be an easy exercise for the reader. It is sufficient to substitute (73) back into (72) and to recall the definition (4.18) of the characteristic polynomial of a graph.

Theorem 6.14. *If the relation (73) holds and $w_i = 0$ for all $i \neq r$, then*

$$\varphi(G, W) = \varphi(G) - w_r \varphi(G - v_r) \, . \tag{75}$$

Proof. Without loss of generality we may assume that $r = 1$. Then the characteristic polynomial of (G, W) is given by

$$\det (xI - A - \text{diag} (w_1, 0, 0, \ldots , 0)) \, . \tag{76}$$

Applying Eq. (16) from Appendix 2 we get

$$\varphi(G, W) = \det (xI - A) + \det (\text{diag} (-w_1, x, x, \ldots , x) - A') \tag{77}$$

where the matrix A' coincides with A, except in its first row:

$$A' = \begin{bmatrix} 0 & 0 & 0 & \cdots & 0 \\ a_{21} & a_{22} & a_{23} & \cdots & a_{2n} \\ a_{31} & a_{32} & a_{33} & \cdots & a_{3n} \\ \cdot & \cdot & \cdot & \cdots & \cdot \\ a_{n1} & a_{n2} & a_{n3} & \cdots & a_{nn} \end{bmatrix}. \tag{78}$$

It is immediately seen that if A is the adjacency matrix of G, then A_1

$$A_1 = \begin{bmatrix} a_{22} & a_{23} & \cdots & a_{2n} \\ a_{32} & a_{33} & \cdots & a_{3n} \\ \cdot & \cdot & \cdots & \cdot \\ a_{n2} & a_{n3} & \cdots & a_{nn} \end{bmatrix} \tag{79}$$

is the adjacency matrix of $G - v_1$.

The first term on the right-hand side of (77) is evidently equal to $\varphi(G)$. Expanding the second determinant we obtain

$$\det(\operatorname{diag}(-w_1, x, x, \ldots, x) - A') = -w_1 \det(\operatorname{diag}(x, x, \ldots, x) - A_1)$$

$$= -w_1 \varphi(G - v_1). \tag{80}$$

Hence the right-hand side of (77) is equal to $\varphi(G) - w_1\varphi(G - v_1)$ and Theorem 6.14 has thus been proved. $\square$

Corollary 6.14.1. *If the relation (73) holds and* $w_i = 0$ *for all* $i \neq r$ *and* $i \neq s$, *then*

$$\varphi(G, W) = \varphi(G) - w_r\varphi(G - v_r) - w_s\varphi(G - v_s) + w_r w_s\varphi(G - v_r - v_s). \tag{81}$$

Theorem 6.14 and its corollary can be further generalized [109].

Theorem 6.15. *If the relation (73) holds, then*

$$\varphi(G, W) = \varphi(G) - \sum_r w_r\varphi(G - v_r) + \sum_{r<s} w_r w_s\varphi(G - v_r - v_s)$$

$$- \sum_{r<s<t} w_r w_s w_t\varphi(G - v_r - v_s - v_t) + \ldots. \tag{82}$$

6.5.3 Some Regularities in the Electronic Structure of Heteroconjugated Molecules

In spite of the fact that the π-electron properties of conjugated molecules have been extensively studied by means of graph theory [24, 30, 31], very few general results have been obtained for heteroconjugated systems. We shall quote here an old [79] and a new one [109].

In paragraph 6.3.3 we have shown that the HMO theory predicts a uniform π-electron charge distribution in alternant hydrocarbons. If one heteroatom is introduced into an alternant conjugated system, the charge distribution is no longer uniform. COULSON and LONGUET-HIGGINS discovered the following "*law of alternating polarity*" [79].

Theorem 6.16. *In an alternant conjugated molecule with one heteroatom, all starred atoms have HMO π-electron charges of the same sign and all unstarred of the opposite sign.*

An immediate consequence of Theorem 6.16 is that π-electron charges alternate in sign along any path in an alternant molecule with one heteroatom. A complete proof of this result has been obtained quite recently [105].

The pairing theorem (Theorem 6.9) does not hold for alternant molecules with heteroatoms. There exists, however, an exception [109]. Let G be a bipartite graph and v_r and v_s its two symmetrically equivalent vertices. Choose a weight matrix W of the form (73) such that $w_i = 0$ for $i \neq r$ and $i \neq s$, and $w_r = -w_s$.

Theorem 6.17. *The pairing theorem holds for the weighted graph (G, W).*

Proof. Apply Corollary 6.14.1. Since v_r and v_s are assumed to be symmetrically equivalent, $G - v_r \simeq G - v_s$ and, consequently, $\varphi(G - v_r) = \varphi(G - v_s)$ and $w_r \varphi(G - v_r) + w_s \varphi(G - v_s) = 0$. Therefore,

$$\varphi(G, W) = \varphi(G) - (w_r)^2 \varphi(G - v_r - v_s) \,. \tag{83}$$

Both G and $G - v_r - v_s$ are bipartite graphs and Eq. (48) applies to both of them. Therefore, (G, W) is also of the form (48).

Now we can prove Theorem 6.17 using the same argument as in the first proof of the pairing theorem. $\square$

Theorem 6.17, although analogous to Theorem 6.9, does not have the same consequences concerning the eigenvectors of (G, W).

Part C

Chemistry and Group Theory

The present part of the book deals with group theory, a very powerful tool in the consideration of the symmetry of a molecule. In Chap. 7 the concept of groups will be briefly displayed and exemplified by means of a particular symmetry group (C_{3v}). The other most common symmetry groups and some of their applications are given in Chap. 8. In Chap. 9 an approach to the automorphism groups of simple graphs, useful in the discussion of the symmetry of non-rigid molecules, is outlined. Finally, in Chap. 10 some particular interrelations between these two types of groups will briefly be discussed. The character tables of some point groups and the first six symmetric groups are given in Appendix 5.

Chapter 7

Fundamentals of Group Theory

Groups are sets of elements amended with a combination law that satisfies certain conditions (axioms).

Definition 1: *Let $\{A, B, C, ...\}$ be a set of elements and let $\odot$ symbolize an operation, the effect of which is defined upon the elements, say for example $A \odot B = C$. Such a set together with the operation is called a group G if the following axioms are satisfied:*
[G 1]: The combination of any two elements of the set by the operation defined results in another element of the set:

$$A, B \in G; \quad \text{if} \quad A \odot B = C, \quad \text{then} \quad C \in G. \tag{1}$$

[G 2]: The set G contains a particular element E which satisfies the following relation for all other elements $A \in G$:

$$E \odot A = A \odot E = A. \tag{2}$$

This particular element E is called the identity element of the group G.
[G 3]: All the operations defined conserve the associative law, i.e.:

$$A \odot (B \odot C) = (A \odot B) \odot C = A \odot B \odot C. \tag{3}$$

[G 4]: For every element, $P \in G$, there exists an inverse one, $P^{-1} = Q \in G$, such that

$$Q \odot P = P \odot Q = E. \tag{4}$$

Of course, according to Eq. (2) the identity element is self-inverse.

One should note that every element of the group commutes with its inverse, as expressed by Eq. (4). In general, however, the conservation of the commutative law needs not to hold, thus $A \odot B$ may differ from $B \odot A$.

Definition 2: The group G is said to be *Abelian* if the commutative law is conserved by all pairs of elements:

$$A \odot B = B \odot A. \tag{5}$$

The wide scope of Definition 1 may be illustrated by the following examples. The complete set of positive and negative integers (inclusively zero) forms an infinite group under the operation of addition wherein the zero represents the identity element and $-A$ is the inverse of $+A$. Other examples of infinite groups are the space groups of crystal lattices [1, 48], but they play only a minor role in organic chemistry. An infinite group of particular interest for atomic physics is the group $O(3)$ which possesses the full symmetry of a sphere in three-dimensional space. The symmetry groups of collinear molecules, $C_{\infty v}$ and $D_{\infty h}$, are also infinite (see paragraph 8.2.3).

For the sake of brevity the symbol $\odot$ is usually omitted. Hence instead of $A \odot B$ we shall write AB.

7.1 The Symmetry Group of an Equilateral Triangle

As an example of a symmetry group we consider the operations which transform the corners of an equilateral triangle into themselves (see Fig. 7.1). They are:

(i) the identity operator E which leaves each point unchanged;
(ii) three reflections A, B and C in a plane perpendicular to the plane of the triangle, passing through the points a, b and c, respectively;
(iii) two rotations through $120°$, D and F, where D is the counterclockwise and F the clockwise rotation.

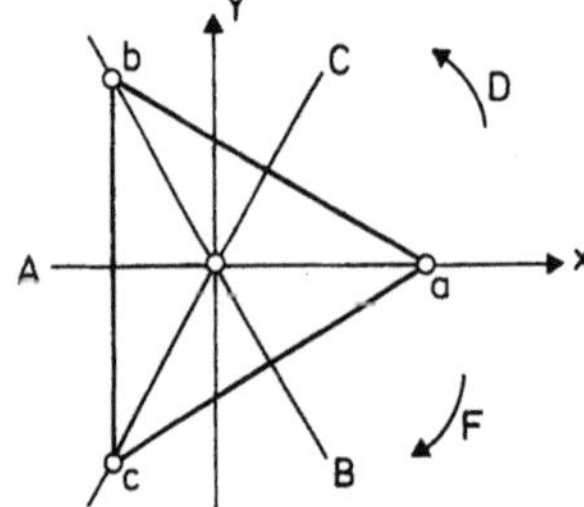

Coordinates:

Point	x	y
a	2ϱ	0
b	$-\varrho$	$\varrho\sqrt{3}$
c	$-\varrho$	$-\varrho\sqrt{3}$

Fig. 7.1. Symmetry operations and coordinates of an equilateral triangle (C_{3v})

That the symmetry operations $\{E, A, B, C, D, F,\}$ form a group is proved by setting up the following *multiplication table*.

$X =$	E	A	B	C	D	F
$Y = E$	E	A	B	C	D	F
A	A	E	D	F	B	C
B	B	F	E	D	C	A
C	C	D	F	E	A	B
D	D	C	A	B	F	E
F	F	B	C	A	E	D

(6)

The entries, $Z \in \{E, A, B, C, D, F\}$, of the above table are obtained as follows: first an operator of the heading line, X, is applied to transform the triangle; onto the result of this an operator, Y, of the heading column is applied. Then by inspection of Fig. 7.1, the operator Z is found which performs the same result, $Z = YX$.

The multiplication table shows that the axioms [G 1], [G 2] and [G 3] are satisfied. Further, in accordance with Eq. (4), the inverse elements (axiom [G 4]) are found as follows:

X	E	A	B	C	D	F
X^{-1}	E	A	B	C	F	D

$$(7)$$

Finally, from the multiplication table one can easily decide whether the group considered is Abelian. Since in the case of an equilateral triangle we find $AB = D$, but $BA = F$, this group is not Abelian.

The results achieved hitherto may be summarized as follows: The symmetry operations of an equilateral triangle form a non-Abelian group $\mathbf{G} = \{E, A, B, C, D, F\}$. The usual notation of the group elements (see Sections 8.1 and 8.2.1) is: the reflections A, B, C are denoted by σ_v, the rotations D and F by C_3 and C_3^2 (note that $C_3^3 = E$), and the group $\mathbf{G}$ itself by $\mathbf{C}_{3v}$.

7.2 Order, Classes and Representations of a Group

Definition 3: The number of elements which form a group is called the *order* of the group.

According to this definition the group $\mathbf{C}_{3v}$ has order $h = 6$.

The group $\mathbf{C}_{3v}$ consists of three different types of operations: the identity operation E; the reflections σ_v, and the rotations C_3, C_3^2. Each of these three subsets is called a *class*. In simple point groups the classes may be easily found by inspection. In more complicated cases as well as in the case of non-geometric groups one may find the classes by means of the following definition.

Definition 4: Let P be any element of the group $\mathbf{G}$. The subset $\{Q | Q = X^{-1}PX; X \in \mathbf{G}\}$ of the elements of $\mathbf{G}$ is the *class* to which P belongs. The element $Q = X^{-1}PX$ is called the *conjugate* of element P.

For example, using Definition 4 and the multiplication table, Eq. (6), of the group $\mathbf{C}_{3v}$, the classes $\{E\}$, $\{A, B, C\}$ and $\{D, F\}$ are obtained. In order to obtain all the elements of a class, the complete set of group elements $X \in \mathbf{G}$ must be applied.

The multiplication table represents the structure of the group. By chance, one may find a set of elements, e.g. numbers, matrices, etc., which multiply according to this table. This means that if, for example, the numbers α, β, δ are associated to the operators A, B, D of **G** then these numbers must satisfy $\alpha\beta = \delta$ in accordance with (6). Such a set of numbers (or more general: elements), which need not be all different, forms a *representation* Γ of the group. In the case of $\mathbf{C}_{3v}$ we easily find the following sets of numbers:

$$
\begin{array}{ccccccc}
 & E & A & B & C & D & F \\
\Gamma_1: & 1 & 1 & 1 & 1 & 1 & 1 \\
\Gamma_2: & 1 & -1 & -1 & -1 & 1 & 1
\end{array}
\tag{8}
$$

which agree with the multiplication table (6). Thus Γ_1 and Γ_2 are two different representations of the group; obviously they are inequivalent.

Evidently, the number $+1$ associated to each element of the group will always agree with a one-dimensional representation of the group. It is called the *totally symmetric representation*. As has already been done in (8), we will denote this representation by Γ_1. In an alternative convention, the totally symmetric representation is denoted by A and some additional suffixes, depending on the structure of the group (see for that Sect. 8.1); in automorphism groups (see Chap. 9) it is denoted by Γ_0.

Another representation, Γ_3, is presented by the transformation matrices of the coordinates x and y associated with the symmetry operations under the law of matrix multiplication.

$$
\begin{array}{ccccccc}
 & E & A & B & C & D & F \\
\Gamma_3: & E & A & B & C & D & F
\end{array}
\tag{9}
$$

where

$$
E = \begin{pmatrix} 1 & 0 \\ 0 & 1 \end{pmatrix}, \quad
A = \begin{pmatrix} 1 & 0 \\ 0 & -1 \end{pmatrix}, \quad
B = \begin{pmatrix} -\varkappa & -\mu \\ -\mu & \varkappa \end{pmatrix},
$$
$$
C = \begin{pmatrix} -\varkappa & \mu \\ \mu & \varkappa \end{pmatrix}, \quad
D = \begin{pmatrix} -\varkappa & -\mu \\ \mu & -\varkappa \end{pmatrix}, \quad
F = \begin{pmatrix} -\varkappa & \mu \\ -\mu & -\varkappa \end{pmatrix},
\tag{10}
$$

$\varkappa = 1/2$, and $\mu = \sqrt{3}/2$. Obviously, the identity E must be represented by the unit matrix. The matrices D and F are obtained from the general transformation matrix for a rotation through an angle φ (see Eq. (21) of Appendix 1) by inserting the proper values of φ, 120° and 240°, respectively. The matrix A is simply derived by inspection of Fig. 7.1. The reflection in the plane A keeps the x coordinate unchanged but changes the sign of the y coordinate. The matrices B and C were obtained by the matrix multiplications $B = AD$ and $C = AF$ according to the multiplication table (6). Note that those matrices which correspond to the elements of a given class have equal traces. This will be proved below by Eq. (33).

From the matrices given in (10) an infinite number of sets of matrices can be obtained by similarity transformations (see Appendix 1): $E' = S^{-1}ES$, $A' = S^{-1}AS$, $B' = S^{-1}BS$ etc. It can be easily verified that the matrices E', A', B', C', D', F' also form a representation of the group. For instance, $A'B' = S^{-1}ASS^{-1}BS = S^{-1}ABS = S^{-1}DS = D'$.

Thus the matrices E', A', B', C', D', F' form another representation of the group, which is equivalent to that given in (9).

7.3 Reducible and Irreducible Representations

The matrices given in (10) operate on the coordinates of a single point. By their means we may construct the transformation matrices of the triangle within the x,y-plane:

$$E^* = \begin{pmatrix} E & 0 & 0 \\ 0 & E & 0 \\ 0 & 0 & E \end{pmatrix} \quad A^* = \begin{pmatrix} A & 0 & 0 \\ 0 & A & 0 \\ 0 & 0 & A \end{pmatrix} \quad B^* = \begin{pmatrix} B & 0 & 0 \\ 0 & B & 0 \\ 0 & 0 & B \end{pmatrix}$$

$$C^* = \begin{pmatrix} C & 0 & 0 \\ 0 & C & 0 \\ 0 & 0 & C \end{pmatrix} \quad D^* = \begin{pmatrix} D & 0 & 0 \\ 0 & D & 0 \\ 0 & 0 & D \end{pmatrix} \quad F^* = \begin{pmatrix} F & 0 & 0 \\ 0 & F & 0 \\ 0 & 0 & F \end{pmatrix} \tag{11}$$

where 0 stands for a zero matrix of order 2. As it is easily verified these matrices satisfy the multiplication table (6), hence, they also form a representation of the group C_{3v}. But due to the block diagonal form of all these matrices, the representation given by (11) may be reduced to that given by (9). One says that the representation (11) is *reducible*.

Let us assume that we have found a set of matrices $\{E', A', B', C', D', F'\}$ of order n which satisfy the multiplication table (6), e.g.:

$$A'B' = D'. \tag{12}$$

Suppose further that there is a similarity transformation which transforms all the matrices E', A', ... into the block diagonal form

$$A'' = S^{-1}A'S = \begin{pmatrix} A_1'' & 0 & 0 & \cdots \\ 0 & A_2'' & 0 & \cdots \\ 0 & 0 & A_3'' & \cdots \\ & \cdot & \cdot & \cdots \end{pmatrix} \tag{13}$$

where A_j'', $j = 1, 2, 3, \ldots$, is a square matrix of the same order as B_j'', C_j'', etc. Then from the law of matrix multiplication and from Eq. (12) one obtains the following relations:

$$A_1''B_1'' = D_1''$$

$$A_2''B_2'' = D_2'' \tag{14}$$

etc.

and the sets of matrices $\{E_1'', A_1'', B_1'', C_1'', D_1'', F_1''\}$, $\{E_2'', A_2'', B_2'', C_2'', D_2'', F_2''\}$, etc. also form representations of the group. Once again one says that the matrix representation E', A', B', ... is reducible and has been reduced by means of a similarity transformation with the matrix S.

A matrix representation is said to be *irreducible* if it is not possible to find a similarity transformation which reduces to block diagonal form all the matrices of this representation. Evidently, *any one-dimensional representation* and, in particular, *the totally symmetric representation is always irreducible*. A precise criterion for the reducibility of a representation is given in Sect. 7.4.

The representations Γ_1, Γ_2 and Γ_3 given in (8) and (9) are irreducible.

Two irreducible representations are said to be *equivalent* if they differ only by a similarity transformation.

As it will be shown below the representations Γ_1, Γ_2 and Γ_3 are the only non-equivalent irreducible representations of the group C_{3v}.

Without proof, we give now certain general theorems on irreducible representations of groups.

Denote by R a symmetry operation of the group considered. Let $\Gamma_i(R)$ be the matrix corresponding to that symmetry operation in the irreducible representation Γ_i and let $\Gamma_i(R)_{mn}$ denote its element in the m-th row and n-th column. Then let us consider the sums

$$\sum_R \Gamma_i(R)^*_{mn}\, \Gamma_j(R)_{m'n'} \tag{15}$$

where $\Gamma_i(R)^*_{mn}$ denotes the complex conjugate of the number $\Gamma_i(R)_{mn}$. If the group representations considered are real, then $\Gamma_i(R)^*_{mn} = \Gamma_i(R)_{mn}$.

In the case of the group C_{3v} we obtain from the irreducible representations (8) and (9) a series of relations among which the only non-zero sums are

$$\sum_R \Gamma_1(R)^*_{11}\, \Gamma_1(R)_{11} = 6 \tag{16}$$

$$\sum_R \Gamma_2(R)^*_{11}\, \Gamma_2(R)_{11} = 6 \tag{17}$$

$$\sum_R \Gamma_3(R)^*_{11}\, \Gamma_3(R)_{11} = \sum_R \Gamma_3(R)^*_{12}\, \Gamma_3(R)_{12} = \sum_R \Gamma_3(R)^*_{21}\, \Gamma_3(R)_{21}$$

$$= \sum_R \Gamma_3(R)^*_{22}\, \Gamma_3(R)_{22} = 3\,. \tag{18}$$

This result may be generalized to

$$\sum_R \Gamma_i(R)^*_{mn}\, \Gamma_i(R)_{mn} = h/l_i \tag{19}$$

where h is the order of the group and l_i is the *dimension* of the i-th irreducible representation, Γ_i. Taking into account that all the other sums result to zero one may write

$$\sum_R \sqrt{l_i/h}\, \Gamma_i(R)^*_{mn}\, \sqrt{l_j/h}\, \Gamma_j(R)_{m'n'} = \delta_{ij}\delta_{mm'}\delta_{nn'} \tag{20}$$

where δ_{ij} denotes the KRONECKER delta which has only two values, namely, $\delta_{ij} = 1$ if $i = j$ and $\delta_{ij} = 0$ if $i \neq j$.

It has been shown that Eq. (20) holds for the non-equivalent irreducible representations of any group.

The factors of the summands in Eq. (20) are the matrix components $\Gamma_i(R_1)_{mn}$, $\Gamma_i(R_2)_{mn}$, ... , of the h symmetry operators forming the group; they may be regarded as the components of an h-dimensional vector defined for a certain irreducible representation, Γ_i, and a certain element (mn) of its matrices. As Eq. (20) shows all these vectors are orthogonal to each other. Since in an n-dimensional vector space at most n orthogonal vectors may be constructed, one can conclude from Eq. (20) that there are exactly h such h-dimensional vectors. On the other hand in each irreducible representation Γ_i of the dimension l_i we have l_i^2 different elements (mn) of the matrices, thus there are l_i^2 such h-dimensional vectors associated with Γ_i. Summing over all irreducible representations one obtains

$$l_1^2 + l_2^2 + l_3^2 + \ldots = h \, . \tag{21}$$

Since this result is a consequence of Eq. (20) which has general validity, the result given in Eq. (21) is true for the non-equivalent irreducible representations of any group.

From Eq. (21) one may conclude that the irreducible representations, Γ_1, Γ_2 and Γ_3 given in (8) and (9) are the only non-equivalent irreducible representations of the group C_{3v}.

7.4 Characters and Reduction of a Reducible Representation

The traces of the matrices which form a given irreducible representation are invariant under similarity transformations (see Appendix 1); they are called the *characters* of the symmetry operations in the given irreducible representation. Let $\chi_i(R)$ denote the character of the symmetry operation R in Γ_i. Then we have

$$\chi_i(R) = \sum_m \Gamma_i(R)_{mm} \, . \tag{22}$$

Since the identity operation E is always represented by a unit matrix E, from Eq. (22) one obtains

$$\chi_i(E) = l_i \, . \tag{23}$$

Consequently, Eq. (21) can be written as

$$h = \Sigma \, [\chi_i(E)]^2 \tag{24}$$

where the sum runs over all irreducible representations.

As a kind of counterpart to the table of irreducible representations the *character table of the group* can be set up. In the case of C_{3v}, for example, one obtains from (8) and (9):

	E	σ_v	σ_v	σ_v	C_3	C_3^2
		A	B	C	D	F
$A_1 = \Gamma_1$	1	1	1	1	1	1
$A_2 = \Gamma_2$	1	-1	-1	-1	1	1
$E = \Gamma_3$	2	0	0	0	-1	-1

$$(25)$$

In the following we shall see that a one-to-one correspondence exists between the irreducible representations and group characters. Hence, it is sufficient to use either the former or the latter.

In order to illustrate the meaning of group characters, some internal motions of ammonia, NH_3, are considered. NH_3 has C_{3v}-symmetry: the equilibrium positions of the hydrogens form an equilateral triangle and the nitrogen is located above its center representing the apex of a regular triangular pyramid. The elongation vectors of the N-H-bonds are denoted by r_a, r_b and r_c, respectively (see Fig. 7.2a); they transform under the operations of C_{3v} as follows:

	E	σ_v	σ_v	σ_v	C_3	C_3^2
		A	B	C	D	F
r_a	r_a	r_a	r_c	r_b	r_b	r_c
r_b	r_b	r_c	r_b	r_a	r_c	r_a
r_c	r_c	r_b	r_a	r_c	r_a	r_b

$$(26)$$

From these transformation properties and the character table of C_{3v}, Eq. (25), the so-called symmetry-adapted functions are obtained by means of

$$\psi_j = \sum_R \chi_j(R) (R\alpha) \tag{27}$$

where α is an element of the basis, $\alpha \in \{r_a, r_b, r_c\}$ and $R\alpha$ is taken from (26). For $\Gamma_1 = A_1$ one obtains

$$a_1 = r_a + r_b + r_c, \tag{28}$$

shown in Fig. 7.2a; this function is related to the symmetric stretching mode of NH_3. The function constructed for $\Gamma_2 = A_2$ vanishes identically. The following three functions obtained for $\Gamma_3 = E$

$$e_1 = 2r_a - r_b - r_c,$$

$$e_2 = 2r_b - r_c - r_a, \tag{29}$$

$$e_3 = 2r_c - r_a - r_b,$$

are shown in Fig. 7.2b, c, and d. They are *equivalent* and differ only in their spatial orientation. Consequently, their motions have the same energy and, therefore they are called *degenerate*. Since $e_1 + e_2 + e_3 = 0$ they are linearly dependent, but they may be transformed into a pair of independent degenerate functions, $e_1' = e_1$ and $e_2' = e_2 - e_3 = r_b - r_c$ (Fig. 7.2e) which are related to the asymmetric stretching modes of NH_3.

The orthogonal functions e_1' and e_2' span an invariant subspace. They transform under the operations of the group according to

$$Re_j' = \sum_{k=1}^{2} \Gamma_3(R)_{jk}\, e_k' \, .$$

This result may be generalized for an irreducible representation Γ_i of dimension l_i as follows: Let the orthogonalized equivalent functions of Γ_i be denoted by ψ_j, $j = 1, 2, \ldots, l_i$; then

$$R\psi_j = \sum_{k=1}^{l_i} \Gamma_i(R)_{jk}\psi_k \tag{30}$$

is satisfied by each element R of the group.

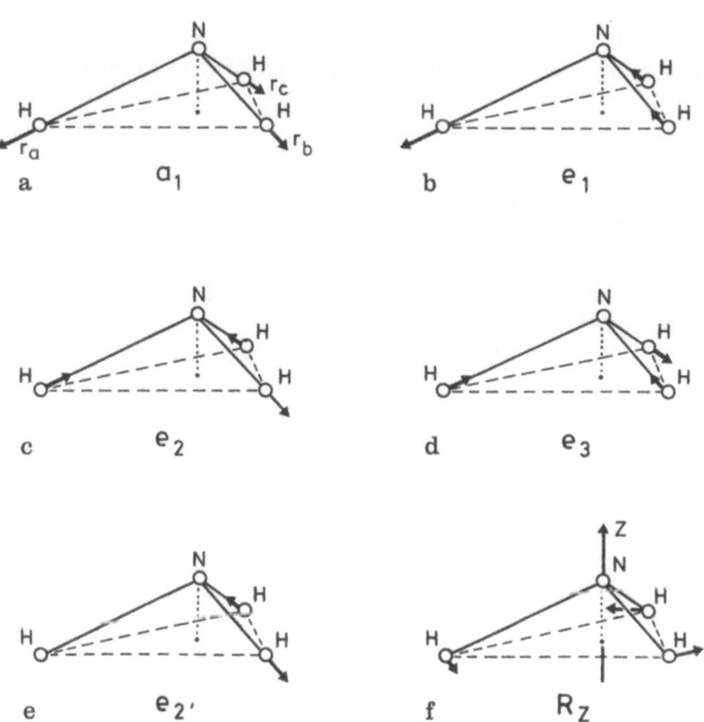

Fig. 7.2. Symmetry-adapted functions in NH_3 molecule: **a** symmetric stretching mode; **b** to **e** asymmetric stretching modes of the NH bonds; **f** rotation about the z-axis

Definition 5: A function ψ_j is said to transform according to a given irreducible representation Γ_i if Eq. (30) is satisfied by each element of the group.

In the case of a one-dimensional representation Eq. (30) reduces to

$$R\psi_j = \chi_i(R)\,\psi_j \, . \tag{31}$$

In Fig. 7.2f each hydrogen is associated with an elongation vector, $\mathbf{t}_a$, $\mathbf{t}_b$ and $\mathbf{t}_c$, respectively, which belong to the xy-plane and have counterclockwise tangential directions with respect to the circle passing through the hydrogens. As is easily verified, the function

$$\mathbf{R}_z = \mathbf{t}_a + \mathbf{t}_b + \mathbf{t}_c \tag{32}$$

belongs to $\Gamma_2 = A_2$ and represents a rotation of the entire ammonia molecule around the z-axis.

The normal coordinates of a molecule [66] are constructed in a similar manner (see paragraph 8.4.3).

As shown in Appendix 1 the trace (character) of a matrix is invariant under similarity transformations. Here we shall consider the case when the matrix $\mathbf{X}$ used for the similarity transformation belongs to a representation of the group. Let $\mathbf{P}$ and $\mathbf{Q}$ be two different elements, $\mathbf{P} \neq \mathbf{Q}$, of a matrix representation of $\mathbf{G}$, and assume that $\mathbf{P}$ and $\mathbf{Q}$ belong to the same class of $\mathbf{G}$; then, as a consequence of Definition 4, in this representation there is at least one element $\mathbf{X} \in \mathbf{G}$ which satisfies the equality $\mathbf{Q} = \mathbf{X}^{-1}\mathbf{P}\mathbf{X}$. In view of Eq. (22) one obtains the character of $\mathbf{Q}$ as follows:

$$\chi(\mathbf{Q}) = \sum_j Q_{jj} = \sum_j \sum_k \sum_l X_{jk}^{-1} P_{kl} X_{lj} = \sum_l \sum_j \sum_k X_{lj} X_{jk}^{-1} P_{kl}$$
$$= \sum_l \sum_k \delta_{lk} P_{kl} = \sum_k P_{kk} = \chi(\mathbf{P}) \, . \tag{33}$$

This equation means that if two operations belong to the same class they have the same characters in a given irreducible representation.

In order to apply Eq. (20) to the characters one has to put $m = n$ and $m' = n'$. This results in

$$\sum_R \sqrt{l_i l_j}\; \Gamma_i(R)_{mm}^{*} \Gamma_j(R)_{m'm'} = h\delta_{ij}\delta_{mm'} \tag{34}$$

where the indices m and m' run as follows: $1 \leq m \leq l_i$, $1 \leq m' \leq l_j$; $l_i \leq l_j$. Summing Eq. (34) over all values of m and m' we obtain with regard to Eq. (22) the following expression

$$\sum_R \chi_i(R)^{*}\chi_j(R) = h\delta_{ij} \, . \tag{35}$$

From Eq. (35) one may conclude:

[C 1]: The characters of the irreducible representations form orthogonal vectors.

[C 2]: Two non-equivalent irreducible representations, i.e. $i \neq j$, have different character systems and two irreducible representations having the same character system are equivalent.

[C 3]: The sum of the squares of the characters in a given irreducible representation equals the order h of the group.

[C 4]: The sum of the characters in a given irreducible representation equals zero if the representation considered is not the totally symmetric one.

Due to the general validity of Eq. (20) the statements above are true for the characters of the non-equivalent irreducible representations of any group.

The statement [C 3] may be used as a criterion for the *reducibility of a representation*: Suppose that for a given representation Γ_i the expression on the left-hand side of Eq. (35) has a value greater than h. Then the representation considered is reducible.

The statement [C 4] is obtained by means of Eq. (35) if Γ_i is supposed to be the totally symmetric representation, i.e. $\chi_i(R) = 1$ for all $R \in G$.

Let k denote the number of classes of a given group, g_ϱ the number of group elements in class ϱ, and finally R_ϱ any one of these elements. Since according to Eq. (33) all elements belonging to the same class have the same character, the left-hand side of Eq. (35) may be rewritten as a summation over the classes as

$$\sum_{\varrho=1}^{k} g_\varrho \chi_i(R_\varrho)^* \chi_j(R_\varrho) = h\delta_{ij} \, . \tag{36}$$

With regard to this expression the normalized characters $\sqrt{g_\varrho/h}\,\chi_i(R_\varrho)$ may be considered as the components of a set of orthonormal k-dimensional vectors. Since there can be at most k such vectors, each one of which is associated with a particular irreducible representation, one can conclude [36] from Eq. (36): The number of irreducible representations equals the number of classes.

Any matrix representation of a group must be either irreducible or reducible. In the latter case it is always a particular combination of irreducible representations in the sense of Eq. (13). As shown by Eq. (13) a reducible representation can be reduced to its irreducible representations by means of an appropriate similarity transformation. Since the characters are defined as the traces of matrices and the traces are invariant under similarity transformations, we can express the character of a matrix R of the reducible representation as

$$\chi(R) = \sum_{j=1}^{k} a_j \chi_j(R) \tag{37}$$

where a_j denotes how many times the j-th irreducible representation occurs in the considered reducible one. Applying Eq. (35) we have

$$\sum_{R} \chi(R)^* \chi_i(R) = \sum_{R} \sum_{j} a_j \chi_j(R)^* \chi_i(R) = ha_i \, . \tag{38}$$

Thus, the number of times the irreducible representation Γ_i occurs in the reducible representations is

$$a_i = \frac{1}{h} \sum_R \chi(R)\, \chi_i(R). \tag{39}$$

Applying group theory to chemical problems, it is more convenient to use group characters than matrices of irreducible representations. Due to the one-to-one correspondence between the system of the characters and that of the irreducible representations of a group both treatments are equivalent. Note however that for a multidimensional irreducible representation there is a unique system of characters whereas an infinite number of matrix representations may be constructed. From this the high significance of group characters can be understood.

The group character table of any group can be constructed from the relations given above.

For convenience we summarize the rules derived above which hold for any finite group:

Rule 1: The number of irreducible representations and the number of classes of the group are equal.

Rule 2: The sum of the squares of the dimensions of the irreducible representations of a group equals the order of the group; see Eqs. (21) and (24).

Rule 3: The characters of an irreducible representation form a vector; the vectors corresponding to non-equivalent irreducible representations are orthogonal; see [C 1] and [C 2].

Rule 4: The sum of the squares of the characters of a given irreducible representation equals the order of the group; see [C 3].

Rule 5: The sum of the characters of a given irreducible representation equals the order of the group if the irreducible representation is the totally symmetric one, but it equals zero otherwise; see [C 4].

Rule 6: Two elements belonging to the same class have the same character in a given irreducible representation; see Eq. (33).

Rule 7: In each group there is one and only one totally symmetric representation; its dimension is 1.

7.5 Subgroups and Sidegroups — Products of Groups

Out of the multiplication table given in (6) one may extract the following part

	E	D	F
E	E	D	F
D	D	F	E
F	F	E	D

$$\tag{40}$$

From this and Definition 1 one easily concludes that the subset $\mathbf{G}' = \{E, D, F\}$ represents a group of order $h' = 3$. Such a group $\mathbf{G}'$ of order h' which is contained in a group $\mathbf{G}$ of order h, $h' < h$, is called a *subgroup* $\mathbf{G}'$ of the group $\mathbf{G}$ while $\mathbf{G}$ is the *supergroup* of $\mathbf{G}'$; these relations are indicated by $\mathbf{G}' \subset \mathbf{G}$. The complement $\bar{\mathbf{G}}'$ of the subgroup $\mathbf{G}'$ with respect to the group $\mathbf{G}$ is called the *sidegroup* of $\mathbf{G}'$. (In our example, $\bar{\mathbf{G}}' = \{A, B, C\}$.) Since a sidegroup does not contain the identity element, it cannot satisfy the axiom [G 2] and, consequently, the elements of a sidegroup never form a group on their own.

Note that the identity element itself also forms a group $\{E\}$ of order one. Hence, each group $\mathbf{G}$ contains two *trivial subgroups*, namely the group $\mathbf{G}$ itself and the group $\{E\}$ formed by the identity element only. Certainly, the non-trivial subgroups are more interesting than the trivial ones. Synonymous terms for trivial and non-trivial subgroups are *proper* and *improper subgroups*, respectively.

By inspection of the multiplication table of the group $\{E, A, B, C, D, F\}$ given in (6) one may find several non-trivial subgroups. In addition to the already mentioned subgroup $\{E, D, F\}$, we also have $\{E, A\}$, $\{E, B\}$ and $\{E, C\}$. These latter groups are *equivalent* because each consists of the identity element and one reflection; they are said to be *isomorphic* (see Sect. 7.7).

Note that all proper subgroups mentioned above are Abelian.

We will now show that the set of elements $\{E, A, B, C, D, F\}$ of the group $\mathbf{G}$ is the cartesian set-product of the elements of two smaller groups, $\mathbf{G}_1 = \{E, A\}$ and $\mathbf{G}_2 = \{E, D, F\}$. Applying the multiplication law of sets one obtains

$$\{E, A\} \otimes \{E, D, F\} = \{EE, ED, EF, AE, AD, AF\} . \tag{41}$$

According to Eq. (2) we have

$$EE = E, \qquad ED = D, \qquad EF = F, \qquad AE = A . \tag{42}$$

With the assignments

$$AD = B \quad \text{and} \quad AF = C , \tag{43}$$

which agree with the notations of (6), one obtains indeed

$$\{E, A\} \otimes \{E, D, F\} = \{E, D, F, A, B, C\} . \tag{44}$$

This relation may be generally expressed by

$$\mathbf{G}_1 \odot \mathbf{G}_2 = \mathbf{G} \tag{45}$$

where $\odot$ stands either for $\oplus$ or for $\oslash$. In the first case ($\oplus$) one says[1] that the group $\mathbf{G}$ is the *direct product* of the groups $\mathbf{G}_1$ and $\mathbf{G}_2$; in the second case ($\oslash$) the product defined by Eq. (45) is termed the *semidirect product* [1] of $\mathbf{G}_1$ and $\mathbf{G}_2$. The difference

---

[1] In the theory of symmetry groups (see Chap. 8) the symbol $\otimes$ or $\times$ is usually used for direct products. In the field of permutation groups, however, to which the automorphism groups of graphs (discussed in Chap. 9) belong, direct products are indicated by $\oplus$ while the symbol $\otimes$ is used for cartesian products of permutation groups. Although cartesian group products are not discussed within this book, for the sake of the uniformity of notation we use $\oplus$ for indicating group direct products even in the case of symmetry groups. Semidirect products are indicated by the symbol $\oslash$ as usual [1].

between these two types of group products is the following. Let R_1 and R_2 be arbitrary elements of the groups $\mathbf{G}_1$ and $\mathbf{G}_2$, respectively. If the commutation law $R_1 R_2 = R_2 R_1$ holds for all the combinations of any $R_1 \in \mathbf{G}_1$ with any $R_2 \in \mathbf{G}_2$ then the group $\mathbf{G}$ in Eq. (45) is the direct product. Otherwise $\mathbf{G}$ is the semidirect product of $\mathbf{G}_1$ and $\mathbf{G}_2$. For more details see Chaps. 4 and 19 of [1].

As seen from the multiplication table, Eq. (6), $AD = B$ but $DA = C$, i.e. $AD \neq DA$; hence, the group $\{E, D, F, A, B, C\}$ is the semidirect product of $\{E, A\}$ and $\{E, D, F\}$.

If the factor groups, $\mathbf{G}_1$ and $\mathbf{G}_2$ have the orders $h(\mathbf{G}_1) = h_1$ and $h(\mathbf{G}_2) = h_2$, respectively, the order of the resulting group $\mathbf{G} = \mathbf{G}_1 \odot \mathbf{G}_2$ is the product of these orders:

$$h(\mathbf{G}_1 \odot \mathbf{G}_2) = h(\mathbf{G}_1)\, h(\mathbf{G}_2) \,. \tag{46}$$

Equation (46) is related to LAGRANGE's theorem which states that the order of a proper subgroup is a divisor of the order of the group. Its immediate consequence is the following result.

Rule 8: If the order of a group is prime, then this group does not contain any non-trivial subgroup. Such groups are called primitive groups; they are Abelian.

When a group $\mathbf{G}$ is generated as a product according to Eq. (45), then the number $k(\mathbf{G})$ of its classes may be equal to or less than the product of the number of classes of the factor groups, i.e. $k(\mathbf{G}) \leq k(\mathbf{G}_1) \cdot k(\mathbf{G}_2)$. Equality holds for direct products. The consequences of this concerning the construction of the character table of a group are discussed in paragraph 8.2.1.

Note that the direct product of two Abelian groups is Abelian, but the semidirect product is not. This latter may be verified in the case of Eq. (44).

Because $D^2 = F$ and $D^3 = E$, the group $\{E, D, F\}$ may be generated by the powers of D. Hence the symmetry element D is the *generator* [50] of the group $\{E, D, F\}$. The generator of $\{E, A\}$ is obviously A. From Eq. (44) it is seen that the group $\{E, A, B, C, D, F\}$ is generated by A and D.

7.6 Abelian Groups

According to Definition 2 the elements of an Abelian group commute with each other. This has a very noticeable consequence.

Rule 9: In Abelian groups each element forms a class on its own.

Really, let P and Q be two elements of a given class. Then by Definition 4, $Q = X^{-1}PX$ for some $X \in \mathbf{G}$. Now since $PX = XP$ and $X^{-1}X = E$ we always obtain

$$Q = X^{-1}PX = X^{-1}XP = EP = P \,. \tag{47}$$

Since the group consists of h elements, according to Rule 1, we also have h irreducible representations. From Rule 2 and Eq. (21) one immediately obtains that each of the irreducible representations has dimension 1.

Rule 10: All irreducible representations of an Abelian group are one-dimensional.

We will now consider in some detail an Abelian group of order $n > 2$ generated by the powers of a particular symmetry element A. This is the cyclic group of order n. Its elements are A^k, $1 \leq k \leq n - 1$, and $A^n = A^0 = E$. The law of multiplication within this group is given by

$$A^k A^l = A^{k+l} . \tag{48}$$

Because of

$$A^n = A^0 = E \tag{49}$$

the inverse of the element A^k is given by

$$(A^k)^{-1} = A^{-k} = A^{n-k} . \tag{50}$$

In order to find the characters of the elements of this group one must keep in mind that they must multiply according to Eq. (48) and must satisfy Eq. (49). Let α_j be the character of A in the irreducible representation Γ_j, i.e. $\chi_j(A) = \alpha_j$, then we have

$$\chi_j(A^k) = \alpha_j^k \tag{51}$$

and bearing in mind Eq. (23),

$$\chi_j(A^n) = \alpha_j^n = \chi_j(E) = 1 \tag{52}$$

since all irreducible representations are 1-dimensional. From Eq. (52) one finds

$$\alpha_j = \sqrt[n]{1} = e^{2j\pi i/n} = \omega^j \tag{53}$$

for $j = 0, 1, 2, \ldots , n - 1$, where $\omega = \exp(2\pi i/n)$, $\omega^* = \exp(-2\pi i/n)$ and $i = \sqrt{-1}$. Thus, the character table of the cyclic group of order n [50] reads as follows:

Irreducible representation	E	A	A^2	...	A^{n-2}	A^{n-1}	
Γ_0	1	1	1	...	1	1	
$\{\ \Gamma_1$	1	ω	ω^2	...	ω^{*2}	ω^*	
$\{\ \Gamma_{n-1}$	1	ω^*	ω^{*2}	...	ω^2	ω	
$\{\ \Gamma_2$	1	ω^2	ω^4	...	ω^{*4}	ω^{*2}	(54)
$\{\ \Gamma_{n-2}$	1	ω^{*2}	ω^{*4}	...	ω^4	ω^2	
etc.	...	...	...	...	...	...	
$(\Gamma_{n/2})$	(1	-1	1	...	1	-1)	

As a consequence of Eq. (53), in cyclic groups there are always pairs of irreducible representations whose characters are pairwise complex conjugate; for examples see

Γ_1 and Γ_{n-1}, Γ_2 and Γ_{n-2}, etc., in (54). They are called *complex conjugate representations*. The totally symmetric representation Γ_0 cannot be paired in such a manner. In the case of even n another representation, $\Gamma_{n/2}$, is present which also cannot be paired. Due to Eq. (53) its character is $\chi_{n/2}(A) = -1$, hence, it is called antisymmetric with respect to A.

The character table given by (54) is typical for cyclic groups. The characters are in general complex numbers as defined by Eq. (53). Note that all cyclic groups are Abelian, but not all Abelian groups are cyclic groups.

Abelian groups have some importance. As already mentioned in Sect. 7.5 all primitive groups of prime order ($h = 2, 3, 5, 7$, etc.) are Abelian; just these groups are most frequently used to generate groups of higher order by means of group products.

7.7 Abstract Groups and Group Isomorphism

The group $\{E, A, B, C, D, F\}$, whose multiplication table is given by (6), has been derived by considering the symmetry of an equilateral triangle. This means that each element of the group has been uniquely associated with a particular well-defined geometric operation. However, even if we did not know the origin of the table (6), an inspection of it would show that (*i*) the six abstract elements A, B, C, D, E and F form a group, say $\mathbf{G}$, under the law of non-commutative multiplication, (*ii*) E represents the identity and (*iii*) A, B and C form one, D and F another class. From this, Rule 1, Eqs. (20) and (21), the character table given in (25) would be obtained. From all these we learn that completely abstract elements may also form a group.

Definition 6: An *abstract group* is formed from abstract elements under a defined law for the combination of pairs of elements.

The group

$$\mathbf{G} = \{E, A, B, C, D, F\} \tag{55}$$

can be understood as such an abstract group as well as the Abelian group given in (54).

With the usual notation (see paragraph 8.2.1) the symmetry group of an equilateral triangle may be denoted as

$$\mathbf{C}_{3v} = \{E, \sigma_v^a, \sigma_v^b, \sigma_v^c, C_3, C_3^2\} \,. \tag{56}$$

Obviously, the groups $\mathbf{G}$ and $\mathbf{C}_{3v}$ have exactly the same structure. Further, each element of the group corresponds uniquely to one and only one element of the other group. Two groups which are related in this manner are said to be *isomorphic*. If two groups $\mathbf{G}_1$ and $\mathbf{G}_2$ are isomorphic, then we shall write $\mathbf{G}_1 \simeq \mathbf{G}_2$.

As seen in paragraph 8.2.1, $\mathbf{D}_3$ is another symmetry group isomorphic with $\mathbf{G}$. Since $\mathbf{C}_{3v}$ and $\mathbf{D}_3$ are isomorphic with the same abstract group, they are said to be isomorphic too. It can be shown that $\mathbf{C}_{nv}$ and $\mathbf{D}_n$ are isomorphic for all values of n. Another example for such a general isomorphism is: $\mathbf{D}_{2n}$ and $\mathbf{D}_{nd}$ are isomorphic if n is even, whereas $\mathbf{D}_{2n}$ and $\mathbf{D}_{nh}$ are isomorphic if n is odd.

Chapter 8

Symmetry Groups

8.1 Notation of Symmetry Elements and Representations

In 3-dimensional space only the few following types of symmetry operations are feasible in order to transform a symmetrical molecule into itself:

E ... the *identity operation* which keeps all the points unchanged.

C_n ... a *rotation* about an axis of symmetry by the angle $2\pi/n$ where n denotes any natural number; C_n is called an *n-fold axis of rotation*. C_n and C_n^{n-1} differ only in the direction of the rotation. The value $n = 1$ indicates that no symmetry axis exists, i.e. $C_1 = E$; the values $n \geq 7$ are only occasionally realized except C_∞ which coincides with the molecular axis in collinear molecules. Among several axes of symmetry that one which has the largest value of n is called the *principal axis*. The rotations obey the following relations:

$$C_n^k C_n^l = C_n^{k+l}; \qquad C_{km}^k = C_m; \qquad C_n^n = E. \tag{1}$$

σ_h ... a *reflection* in a symmetry plane which is perpendicular to the principal axis.

σ_v ... a *reflection* in a symmetry plane which contains the principal axis.

σ_d ... a *reflection* in a plane of symmetry which contains the principal axis as well as bisects the angle between two C_2 axes which are perpendicular to the principal axis.

The reflections obey the following relations:

$$\sigma_h^2 = E; \qquad \sigma_v^2 = E; \qquad \sigma_d^2 = E. \tag{2}$$

S_n ... a *rotation* about an axis of symmetry by the angle $2\pi/n$ *followed by a reflection* in a plane perpendicular to the axis or vice versa. Thus one has

$$S_n = \sigma_h C_n = C_n \sigma_h. \tag{3}$$

S_n is called an *n-fold alternating axis* or an *n-fold rotation-reflection axis*. S_n and S_n^{n-1} differ only with respect to the direction of the rotation. It is easily verified that $\sigma_h C_2 = S_2$ is identical with the inversion i on a symmetry center (see below). The following relations are obeyed:

$$S_n^{2m} = C_n^{2m}, \qquad 2m < n;$$
$$S_{2n+1}^{2n+1} = \sigma_h, \qquad S_{2n+1}^{4n+2} = E;$$
$$S_{4n+2}^{2n+1} = S_2 = i, \qquad S_{4n}^{2n} = C_2, \qquad S_{2n}^{2n} = E. \tag{4}$$

i ...   the *inversion* in a center of symmetry. Obviously, the following relations hold:

$$i = S_2 = \sigma_h C_2 = C_2 \sigma_h, \qquad i^2 = E. \tag{5}$$

Equations (1) to (5) are useful for the construction of the multiplication table of a given group as well as for the generation of a group by means of the group product of two groups.

By convention, the principal axis of symmetry is always identified as the z-axis of the coordinate system; if the group considered contains a further rotation axis perpendicular to the principal axis then it is identified as the x-axis.

The one-dimensional representations of finite symmetry groups are denoted by A or B, the two-dimensional ones by E, three-dimensional ones by F or T, four-dimensional ones by G, etc., now in alphabetic order (MULLIKEN's notation). A and B distinguish the representations which are, respectively, symmetric or antisymmetric with respect to the rotations about the principal axis of symmetry. These rotations always form an Abelian subgroup isomorphic with the cyclic group (see Sect. 7.6). Within this subgroup, A and B correspond to Γ_0 and $\Gamma_{n/2}$, respectively, as given by Eq. (7.54).

Several representations of the same type are distinguished by subscripts (see for example C_{nv} in the paragraph 8.2.1): A_1 and A_2 are respectively symmetric and antisymmetric with respect to the sidegroup.

The subscripts g ("gerade" = even) and u ("ungerade" = odd) indicate that the representation is symmetric (g) or antisymmetric (u) with respect to the center of symmetry (see for examples $C_{(2n)h}$ in paragraph 8.2.1). If the group considered does not contain a center of symmetry, but possesses a reflection plane, σ_h, perpendicular to the principal axis, then the representations symmetric under σ_h are primed whereas the antisymmetric ones are doubly primed (see for examples $C_{(2n+1)h}$ in paragraph 8.2.1).

According to these conventions, the totally symmetric representation is denoted by one of the following symbols: A, A_1, A_{1g}, A' or A_1'.

In infinite symmetry groups, $C_{\infty v}$ and $D_{\infty h}$, a special notation is used namely, $A_1 = \Sigma^+$, $A_2 = \Sigma^-$, $E_1 = \Pi$, $E_2 = \Delta$, $E_3 = \Phi$, etc.

8.2 Some Symmetry Groups

The groups used to describe the geometric symmetry of finite molecules are point groups. A symmetry group is called a *point group* if all its operations leave one point unchanged. This point coincides with the center of inversion, provided such a center exists. Symmetry groups are usually denoted by boldface letters.

For convenience we shall divide the symmetry groups into the following three classes:

(1) Finite *rotation groups* possessing a single principal axis of symmetry;

(2) *Groups with more than one n-fold axis, n > 2*;

(3) *Groups of collinear molecules.*

In the following the groups most frequently occurring in organic chemistry will be described. The corresponding character tables are collected in Appendix 5.

8.2.1 Rotation Groups

All the rotation groups possess a principal axis of symmetry, C_n, which is usually identified as the z-axis of the coordinate system. The following types of groups belong to this class:

1. *The groups* C_n: These groups consist only of all powers of the rotation C_n. Hence, their order is $h(C_n) = n$. The group C_n is Abelian, and is isomorphic with the cyclic group described in Sect. 7.6. The group C_1 consists only of the identity element, $C_1 = \{E\}$; all molecules without any symmetry belong to C_1. If n is not prime but may be represented by a product of prime numbers, e.g. $n = abc$, then C_a, C_b, C_c, C_{ab}, C_{ac} and C_{bc} are proper subgroups of C_n.

In Appendix 5 the character tables of these groups are given for $n \leqq 6$. As already mentioned, the representations A and B correspond to Γ_0 and $\Gamma_{n/2}$, respectively, of (7.54). The pairs Γ_j and Γ_{n-j}, $0 < j < n/2 > 1$, of the table (7.54) are denoted as components of doubly degenerate representations. The function z always belongs to A; the functions x and y belong to E_1, that is $(x + iy)$ and $(x - iy)$ transform according to the one-dimensional components of E_1.

2. *The groups* C_{nv}: The group C_{nv} may be generated by $C_n \oslash \{E, \sigma_v\}$ and, hence, it consists of all powers of the rotation C_n and n planes of symmetry containing C_n. The order of this group is $h(C_{nv}) = 2n$. Such a group does not exist for $n = 1$. Due to the generation of C_{nv} as a semidirect product of C_n and $\{E, \sigma_v\}$, its side group is produced by $C_n \otimes \{\sigma_v\}$. All these elements are planes of reflection. If n is odd, they form a single class, $\{\sigma_v\}$, of order n, but they split into two classes, $\{\sigma_v\}$ and $\{\sigma_{v'}\}$, of equal order, if n is even.

The character system of the C_n subgroup in C_{nv} is the same as in C_n. But because the character of σ_v in $\{E, \sigma_v\}$ is either $+1$ or -1, each irreducible representation of C_n is split into two representations distinguished by subscripts 1 and 2. In one (subscript 1) the characters of the sidegroup elements $(C_n\sigma_v)$ are identically equal to those of the corresponding subgroup elements, i.e. $\chi_j(C_n\sigma_v) = \chi_j(C_n)$, but in the second one (subscript 2) $\chi_j(C_n\sigma_v) = -\chi_j(C_n)$. The elements C_n and σ_v do not commute. The following equality is readily proved

$$\sigma_v C_n^m = C_n^{n-m} \sigma_v . \tag{6}$$

Since $\sigma_v^2 = E$, from this result one obtains $\sigma_v C_n^m \sigma_v = C_n^{n-m} \sigma_v \sigma_v = C_n^{n-m}$ and $\sigma_v C_n^{n-m} \sigma_v = \sigma_v \sigma_v C_n^m = C_n^m$. This means that in the groups C_{nv}, $n \geqq 3$, the rotation C_n^m and its inverse C_n^{n-m} form a class of order 2 for all m, $1 \leqq m < n/2$. As a consequence of that, the number of classes and irreducible representations is reduced to $k < h = 2n$. In addition to this, the components of C_n with conjugate complex characters collapse into doubly degenerate representations. Such a behaviour is typical for groups generated by a semidirect product. Concerning the characters of the doubly degenerate representations (see Chap. 4 of [1]) note that $\omega^j + \omega^{*j} = 2 \cos j\varphi$, where $\omega = \exp(2\pi i/n)$, $\omega^* = \exp(-2\pi i/n)$ and $\varphi = 2\pi/n$.

3. *The groups* C_{nh}: These groups may be generated by the direct product $C_n \oplus \{E, \sigma_h\}$. Since C_n and σ_h commute as shown by Eq. (3), these groups are Abelian. In view of Eqs. (3) and (4) they contain all powers of the rotation C_n. They further contain σ_h and, if n is even, also i. The other elements of C_{nh} are of S_n type. Consequently, the order of the group is $h(C_{nh}) = 2n$. The group C_{1h} is very often assigned as C_s;

it is the group of all molecules which have no symmetry element other than a plane of symmetry.

The character of σ_h in $\{E, \sigma_h\}$ is either $+1$ or -1. Thus, due to $\mathbf{C}_{nh} = \mathbf{C}_n \oplus \{E, \sigma_h\}$, each representation of $\mathbf{C}_n$ splits into a pair of representations distinguished by the characters of the sidegroup elements $\mathbf{C}_n \otimes \{\sigma_h\}$. These are either the same as for the $\mathbf{C}_n$ subgroup elements or have opposite signs. The representations of such a pair are primed and doubly primed if n is odd or g and u if n is even. Such a splitting of the character system is typical for groups generated as the direct product of two smaller groups, one of which has order 2.

4. *The groups* $\mathbf{S}_n$: These groups consist of all powers of S_n. Hence their order is $h(\mathbf{S}_n) = n$. These groups are Abelian and isomorphic with the cyclic group of order n (see Sect. 7.6). As seen from Eq. (4) they only exist for even n. They always contain $\mathbf{C}_{n/2}$ as a proper subgroup. If $n = 4v + 2$, then according to Eq. (4) the group also contains the inversion i. The group $\mathbf{S}_2$ is usually denoted by $\mathbf{C}_i$.

5. *The groups* $\mathbf{D}_n$: The group $\mathbf{D}_n$ may be generated by $\mathbf{C}_n \oslash \{E, C_2'\}$ and, hence, it consists of the powers of C_n and n 2-fold axes C_2' which are perpendicular to the principal axis C_n. Hence, the order of the group is $h(\mathbf{D}_n) = 2n$. If n is odd all the 2-fold axes are polar and equivalent; if n is even they are apolar but there are $n/2$ two-fold axes of one type, C_2', and $n/2$ two-fold axes of another type, C_2''. The group $\mathbf{D}_2$ is also denoted by $\mathbf{V}$.

It is easy to prove that

$$C_2' C_n^k = C_n^{n-k} C_2'.\tag{7}$$

This has similar consequences as Eq. (6) for the groups $\mathbf{C}_{nv}$ and results in the non-Abelian character of the groups $\mathbf{D}_n$, $n > 2$.

In the group $\mathbf{D}_2$ the three two-fold axes coincide with the axes of the coordinate system and they are equivalent. Consequently, in $\mathbf{D}_2$ there is only one representation of A-type, but three are of B-type. The group $\mathbf{D}_2$ is Abelian.

6. *The groups* $\mathbf{D}_{nd}$: These groups may be generated in accordance with $\mathbf{D}_n \oplus \{E, \sigma_d\}$, hence their orders are $h(\mathbf{D}_{nd}) = 4n$. The product of $C_2' \in \mathbf{D}_n$ and σ_d results in $\{C_2'\} \otimes \{\sigma_d\} = \{S_{2n}\}$, where $\{C_2'\}$, $\{\sigma_d\}$ and $\{S_{2n}\}$ denote the classes containing C_2', σ_d and S_{2n}, respectively. The group $\mathbf{D}_{2d}$ is also denoted by $\mathbf{V}_d$. As a consequence of Eq. (6) the groups $\mathbf{D}_{nd}$ are not Abelian and therefore the representations B_2, $B_3 \in \mathbf{D}_2$ are merged into a doubly degenerate representation $E \in \mathbf{D}_{2d}$.

7. *The groups* $\mathbf{D}_{nh}$: These groups may be generated according to $\mathbf{D}_n \oplus \{E, \sigma_h\}$, where $\{C_n\} \otimes \{\sigma_h\} = \{S_n\}$, $\{C_2'\} \otimes \{\sigma_h\} = \{\sigma_v\}$ and $\{C_2''\} \otimes \{\sigma_h\} = \{\sigma_d\}$. As before, $\{C_n\}$, $\{\sigma_h\}$, $\{S_n\}$ etc. are the classes which contain C_n, σ_h, S_n etc. The order of $\mathbf{D}_{nh}$ is $h(\mathbf{D}_{nh}) = 4n$. The group $\mathbf{D}_{2h}$ is also denoted by $\mathbf{V}_h$; the three symmetry axes as well as the three symmetry planes of $\mathbf{D}_{2h}$ are respectively equivalent. Among the groups $\mathbf{D}_{nh}$ only $\mathbf{D}_{2h}$ is Abelian.

8.2.2 Groups with More than One n-Fold Axis, $n > 2$

Of interest are the cubic groups derived from the tetrahedron, from the octahedron or cube, and the icosahedral groups.

1. *Tetrahedral groups* $\mathbf{T}$, $\mathbf{T}_d$ and $\mathbf{T}_h$: The group $\mathbf{T}$ transforms a regular tetrahedron into itself, the group $\mathbf{T}_d$ is the symmetry group of methane, CH_4, and $\mathbf{T}_h$ is the symmetry group of $[Co(NO_2)_6]^{3-}$.

The three two-fold axes coincide with the axes of the coordinate system; they are equivalent in all these groups and, hence, form a class. There are four three-fold axes coinciding with the cube diagonals a) $x = y = z$, b) $x = -y = -z$, c) $x = y = -z$, and d) $x = -y = z$, respectively; they always belong to one class. The counterclockwise and the clockwise rotations about these axes form different classes in $\mathbf{T}$ and $\mathbf{T}_h$, but a single class in $\mathbf{T}_d$.

2. *Octahedral groups* $\mathbf{O}$ and $\mathbf{O}_h$: The group $\mathbf{O}$ transforms a cube or a regular octahedron into itself while $\mathbf{O}_h$ is the symmetry group of SF_6.

The three two-fold axes, $C_2 = C_4^2$, of these groups coincide with the axes of the coordinate system; the six two-fold axes, C_2', bisect the angles formed from a pair of C_2. The three-fold axes are located as described in the case of $\mathbf{T}$, i.e. they intersect the centers of pairwise parallel triangles of the octahedron.

In Appendix 5 we give only the character table of the group $\mathbf{O}$. The character table of $\mathbf{O}_h$ is simply produced according to $\mathbf{O}_h = \mathbf{O} \oplus \{E, i\}$. In $\mathbf{O}_h$ the functions x, y, z belong to F_{1u}; R_x, R_y, R_z to F_{1g}; $x^2 + y^2 + z^2$ to A_{1g}; $x^2 + y^2 - 2z^2$ and $x^2 - y^2$ to E_g; and finally xy, yz, zx to F_{2g}.

As seen from the character table, $\mathbf{T}$ is a subgroup of $\mathbf{O}$.

3. *Icosahedral groups* $\mathbf{I}$ and $\mathbf{I}_h$: The group $\mathbf{I}$ transforms a regular icosahedron into itself. These groups play some role in boron-organic chemistry. So, for example, the sodium chloride-type lattice of boron carbide, CB_4, is occupied by linear C_3 groups and compact B_{12} groups forming regular icosahedra. Further, the C_{60}-cluster "footballene", obtained recently from graphite in laser experiments, is assumed to exhibit $\mathbf{I}_h$ symmetry.

From all the rotation axes of $\mathbf{I}$ only one C_5 coincides with an axis, say z, of the coordinate system.

In Appendix 5 we only give the character table of the group $\mathbf{I}$ from which that of $\mathbf{I}_h$ is simply obtained according to $\mathbf{I}_h = \mathbf{I} \oplus \{E, i\}$. In $\mathbf{I}_h$ the functions x, y, z belong to F_{1u}; R_x, R_y, R_z to F_{1g}; $x^2 + y^2 + z^2$ to A_{1g}; and all the other bilinear functions as $x^2 + y^2 - 2z^2$, $x^2 - y^2$, xy, yz, zx to the five-fold degenerate representation H_g.

8.2.3 Groups of Collinear Molecules

The principal axis of rotation which coincides with the z-axis is ∞-fold. Hence, the rotation angle φ can assume any value between 0 and 2π. Therefore the order of these groups is infinite. The groups of interest are $\mathbf{C}_{\infty v}$, the symmetry group of collinear molecules without a center of symmetry (like HCN) and $\mathbf{D}_{\infty h}$, the symmetry group of collinear molecules with a center of symmetry (like $HC \equiv CH$).

The counterclockwise and the clockwise rotations about the axis by a given angle φ form a class indicated by $2C_\infty^\varphi$. There is an infinite number of such classes, corresponding to the infinite number of values φ may take. Due to $C_\infty^\pi = C_2$, there is also a two-fold axis coinciding with the z-axis; this C_2 forms a class on its own. The infinite number of symmetry planes σ_v form a single class.

The group $\mathbf{D}_{\infty h}$ may be generated according to $\mathbf{D}_{\infty h} = \mathbf{C}_{\infty v} \oplus \{E, i\}$, where $C_\infty^\varphi i = S_\infty^{2\pi-\varphi}$, $C_2 i = \sigma_h$ and $\sigma_v i = C_2'$. The general structure remains unchanged by the generation procedure. Hence, $\mathbf{D}_{\infty h}$ has the additional classes i, $2S_\infty^\varphi$, σ_h and $\infty C_2'$. Further, each representation of $\mathbf{C}_{\infty v}$ splits into a g and a u representation.

8.3 Transformation Properties and Direct Products of Irreducible Representations

8.3.1 Transformation Properties

First of all, we consider the symmetry transformations of a point which is part of an object exhibiting $\mathbf{D}_{2h}$ symmetry. The symmetry elements of $\mathbf{D}_{2h}$ coincide with the axes, the planes and the origin of the coordinate system (see paragraph 8.2.1).

Definition 1: A point of the space is called an *arbitrary point* if it does not belong to any element of the symmetry group.

Obviously, in $\mathbf{D}_{2h}$ such an arbitrary point, say P_1, has coordinates (x, y, z) with non-zero entries. The various elements of $\mathbf{D}_{2h}$ transform P_1 into 8 points, P_1, P_2, ... , P_8 (see Fig. 8.1), according to which:

$\mathbf{D}_{2h}$	E	C_2^z	C_2^y	C_2^x	i	σ_v^{xy}	σ_v^{zx}	σ_v^{yz}		
P_1	P_1	P_2	P_3	P_4	P_5	P_6	P_7	P_8		(8)
x	x	$-x$	$-x$	x	$-x$	x	x	$-x$	B_{3u}	
y	y	$-y$	y	$-y$	$-y$	y	$-y$	y	B_{2u}	
z	z	z	$-z$	$-z$	$-z$	$-z$	z	z	B_{1u}	

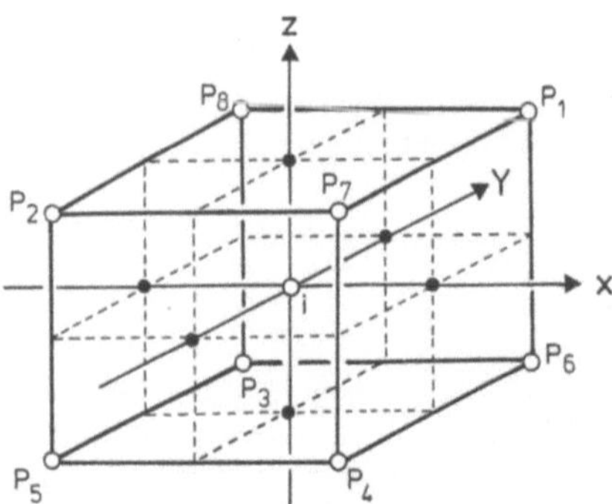

Fig. 8.1. The octuple of points generated from P_1 under the symmetry operations of $\mathbf{D}_{2h}$

The points P_1, P_2, ... , P_8 represent the corners of a rectangular prism. It is easy to see that in the case of $\mathbf{D}_{2h}$ symmetry, all arbitrary points of the space may be grouped into such octuples. As seen from (8), their coordinates are functions of the coordinates of one of these points, say P_1. Thus, in $\mathbf{D}_{2h}$ one may choose the subspace defined by $x > 0$, $y > 0$ and $z > 0$ as the *subspace of independent arbitrary points*. This subspace corresponds to a spherical angle of $4\pi/8$ and it is mapped onto the total space of arbitrary points under the operations of $\mathbf{D}_{2h}$. The total space of all points is obtained by adding the three symmetry planes σ, which do not contain arbitrary points according to Definition 1. In view of the Definition 1 and the group axiom [G 1] these results may be generalized as follows.

Rule 1: Under the operations of a symmetry group of order h an arbitrary point is transformed into exactly h points. The subspace of the independent arbitrary points corresponds to a spherical angle of $4\pi/h$.

Let $f(x, y, z)$ be a symmetry-adapted function which transforms according to the one-dimensional irreducible representation Γ_j of the symmetry group $\mathbf{G}$ of order h, i.e. let $f(x, y, z)$ satisfy Eq. (7.31) for all elements R_k of $\mathbf{G}$, $1 \leq k \leq h$. Let in addition $f(P_1) = f(x_1, y_1, z_1)$ denote the value of this function at an arbitrary point P_1 with the coordinates $(x_1 > 0, y_1 > 0, z_1 > 0)$. Assume that R_k transforms P_1 into P_k, that is $R_k P_1 = P_k$. Then according to Eq. (7.31), the value of $f(x, y, z)$ at the point P_k is given by

$$f(P_k) = \chi_j(R_k) f(P_1) . \tag{9}$$

Summing over all values of k, $1 \leq k \leq h$ and taking into account Rule 5 from Chap. 7, one obtains

$$\sum_{k=1}^{h} f(P_k) = \begin{cases} h f(P_1) & \text{if } \Gamma_j \text{ is totally symmetric} \\ 0 & \text{otherwise.} \end{cases} \tag{10}$$

Equation (10) also holds when $f(x, y, z)$ transforms according to a multidimensional irreducible representation, i.e. $f(x, y, z)$ satisfies Eq. (7.30). In this case Eq. (9) must be appropriately modified.

Note that Eq. (9) holds for any arbitrary point of the space. Thus, for the integral

$$\langle f \rangle = \iiint f(x, y, z) \, dx \, dy \, dz \tag{11}$$

the integration over the complete space may be replaced by an integration over the subspace of independent arbitrary points and the subsequent application of Eq. (10). From this one concludes the following:

Rule 2: An integral of type (11) over the complete space is zero if the integrand does not belong to the totally symmetric representation.

The above result is rather important because integrals of the type (11) frequently occur in various branches of science. Since Rule 2 tells us under what conditions these integrals vanish identically, it is a very powerful tool and gives rise to a series of selection rules.

Deriving Rule 2, it was assumed that the integrand in Eq. (11), $f(x, y, z)$, belongs to a distinct irreducible representation of the group. We now consider another integrand, $F(x, y, z)$, which does not meet that assumption. This means that $F(x, y, z)$ is reducible and may be expanded in terms of symmetry-adapted functions $f_j(x, y, z) \in \Gamma(f_j)$ as

$$F(x, y, z) = \sum_{j=1}^{k} a_j f_j(x, y, z) . \tag{12}$$

The coefficients a_j are determined by means of Eqs. (7.37), (7.38) and (7.39). As easily verified, the integral $\langle F \rangle$ is solely determined by the totally symmetric component of $F(x, y, z)$; for more details see Chaps. 8 and 9 in [11].

With regard to the general validity of Rule 2, we shall now consider the case when the integrand in (11) is a product of symmetry-adapted functions. Suppose that $a_i(x, y, z)$, $i = 1, 2, \ldots$, and $b_j(x, y, z)$, $j = 1, 2, \ldots$, are two sets of functions which form bases for representations of the group. Then the set of functions

$$c_{ij}(x, y, z) = a_i(x, y, z)\, b_j(x, y, z)$$

forms another basis for a representation of the group. The set of functions c_{ij} is called the *direct product* of the sets of functions a_i and b_j. It can be shown (see for example Sect. 10d in [20]) that the characters of c_{ij} are obtained by the application of the following equation for all elements R of the group:

$$\chi(c_{ij}, R) = \chi(a_i, R)\, \chi(b_j, R)\,.$$

This result can be generalized for direct products made up from more than two sets of functions. Let the integrand of (11) be a product of symmetry-adapted functions $g_j(x, y, z) \in \Gamma(g_j), j = 1, 2, 3, \ldots$,

$$f(x, y, z) = \prod_j g_j(x, y, z) \tag{13}$$

then the characters of $f(x, y, z)$ are given as

$$\chi(f, R) = \prod_j \chi(g_j, R)\,. \tag{14}$$

Rule 3: The character of the representation of a direct product is equal to the product of the characters of the individual representations.

The direct product of two irreducible representations will be reducible in general. It may be expanded in terms of the irreducible representations by means of Eqs. (7.37) and (7.39). As easily verified, in the case of the group given by Eq. (7.25) one obtains

$$\Gamma_1 \oplus \Gamma_1 = \Gamma_1\,, \qquad \Gamma_2 \oplus \Gamma_2 = \Gamma_1\,, \qquad \Gamma_3 \oplus \Gamma_3 = \Gamma_1 + \Gamma_2 + \Gamma_3\,,$$
$$\Gamma_1 \oplus \Gamma_2 = \Gamma_2\,, \qquad \Gamma_1 \oplus \Gamma_3 = \Gamma_3\,, \qquad \Gamma_2 \oplus \Gamma_3 = \Gamma_3\,.$$

These results illustrate another general rule.

Rule 4: The direct product of two irreducible representations contains the totally symmetric representation only if the two factors of the direct product are equal.

In the integrals of type (11) which frequently occur in quantum chemistry, the integrand has usually the following form:

$$f(x, y, z) = \psi_j^* \mathscr{F} \psi_k \tag{15}$$

where ψ_j and ψ_k denote the wave functions involved and $\mathscr{F}$ is an operator associated with a particular observable; concerning the construction of such operators see for example Sect. 3d of [20]. If $j = k$ and ψ_j denotes a state function (most frequently of the ground state), then the integral $\langle f \rangle$ is called the expectation value of the observable associated with $\mathscr{F}$.

By means of Rule 3 the symmetry assignments of molecular states may be obtained. For example, the symmetry characters of an electronic state of a molecule, i.e. its electronic configuration, are given by the direct product of those irreducible representations to which the occupied MO's belong. In order to distinguish these different symmetry species, minuscules are used for assigning MO's (e.g. b_{2u}, a''), but capitals for states (e.g. B_{2u}, A''). It can be shown that an electronic configuration belongs to the totally symmetric representation if all its MO's are doubly occupied by electrons or empty and if equivalent degenerate MO's are either all occupied or all empty ("closed shell configuration"). Thus in the case of a closed shell configuration ψ_k, the integrand $\psi_k^* \mathscr{F} \psi_k$ and the operator $\mathscr{F}$ transform according to the same representation. Consequently the integral $\langle f \rangle$ vanishes identically unless $\mathscr{F}$ itself or a component of $\mathscr{F}$ belong to the totally symmetric representation.

Rule 5: For closed shell configurations only those operators have non-zero expectation values which are either totally symmetric or have a totally symmetric component.

Consider now the transformation behaviour of the functions x, y and z under the operations of $\mathbf{D}_{2h}$. A comparison of Eq. (8) with the character table of $\mathbf{D}_{2h}$ (see Appendix 5) shows that in $\mathbf{D}_{2h}$,

$$x \in B_{3u}, \qquad y \in B_{2u}, \qquad z \in B_{1u}.$$

From these variables the following bilinear forms can be constructed: x^2, y^2, z^2, xy, yz and zx. Applying Rule 3 one immediately finds that in $\mathbf{D}_{2h}$,

$$x^2, y^2, z^2 \in A_g, \qquad xy \in B_{1g}, \qquad yz \in B_{3g}, \qquad zx \in B_{2g}. \tag{16}$$

Note that the three p-AO's located at the origin of the coordinate system transform like x, y and z. Similarly, three of the five d-AO's transform like xy, yz and zx. Thus, each of these AO's belongs to a pertinent irreducible representation of $\mathbf{D}_{2h}$. The two remaining d-AO's, namely $d_{x^2-y^2}$ and d_{z^2}, both belong to A_g.

Since $\mathbf{D}_{2h}$ is Abelian, the application of (14) is simple. In the case of $\mathbf{C}_{3v}$ (see (7.25) and paragraph 8.2.1), one finds that $z \in A_1$ whereas x and y form a 2-dimensional irreducible representation E. From this result and Rule 3 one simply concludes that $z^2 \in A$ and yz, $zx \in E$. For the bilinear forms x^2, y^2 and xy from Eq. (14) one obtains the character of $E \oplus E$, i.e. $\chi(E) = 4$, $\chi(2C_3) = 1$, $\chi(3\sigma_v) = 0$. Obviously, this representation is reducible and by means of (7.39) one finds that for $\mathbf{C}_{3v}$

$$E \oplus E = A_1 + A_2 + E. \tag{17}$$

Consequently, $x^2 + y^2 \in A_1$ and $x^2 - y^2$, $xy \in E$. Note that $x^2 - y^2$ and xy are equivalent functions which differ only in their orientation in the xy-plane by an angle of $45°$; hence they are degenerate.

In all rotation groups with an n-fold principal axis, $n > 2$, the functions x and y as well as $x^2 - y^2$ and xy belong either to a doubly degenerate irreducible representation or (in the case when the rotation group is Abelian) to a pair of complex conjugate representations.

8.3.2 Rules Concerning the Direct Product of Irreducible Representations

The procedure performed above for $E \oplus E$ in C_{3v} may be similarly applied to any direct product of two irreducible representations of an arbitrary group. The results obtained are summarized by the following rules.

(1) *General rules:*

$$A \oplus A = A, \qquad\qquad B \oplus B = A, \qquad\qquad A \oplus B = B;$$
$$A \oplus E = E, \qquad\qquad B \oplus E = E,$$
$$A \oplus E_1 = E_1, \qquad\qquad B \oplus E_1 = E_2,$$
$$A \oplus E_2 = E_2, \qquad\qquad B \oplus E_2 = E_1;$$
$$g \oplus g = g, \qquad\qquad u \oplus u = g, \qquad\qquad g \oplus u = u;$$
$$' \oplus ' = ', \qquad\qquad '' \oplus '' = ', \qquad\qquad ' \oplus '' = ''.$$

(2) *Multiplication of subscripts on A or B*

(2.1) except for D_2 and D_{2h}:

$$1 \oplus 1 = 1, \qquad\qquad 2 \oplus 2 = 1, \qquad\qquad 1 \oplus 2 = 2;$$

(2.2) for D_2 and D_{2h}:

$$1 \oplus 2 = 3 \qquad\qquad 2 \oplus 3 = 1 \qquad\qquad 1 \oplus 3 = 2.$$

(3) *Two-dimensional representations:*

(3.1) for C_3, C_{3v}, C_{3h}, D_3, D_{3d}, D_{3h}, C_6, C_{6v}, C_{6h}, D_6, D_{6h}, S_6, T, T_d, T_h, O, O_h:

$$E_1 \oplus E_1 = E_2 \oplus E_2 = A_1 + A_2 + E_2$$
$$E_1 \oplus E_2 = B_1 + B_2 + E_1;$$

(3.2) for C_4, C_{4v}, C_{4h}, D_{2d}, D_{4h}, S_4:

$$E \oplus E = A_1 + A_2 + B_1 + B_2$$

For groups listed above where the symbols A, B, or E are without subscripts, take $A_1 = A_2 = A$, etc.

(4) *Three-dimensional representations:*

(4.1) for T_d, O, O_h:

$$E \oplus F_1 = E \oplus F_2 = F_1 + F_2;$$
$$F_1 \oplus F_1 = F_2 \oplus F_2 = A_1 + E + F_1 + F_2;$$
$$F_1 \oplus F_2 = A_2 + E + F_1 + F_2;$$

(4.2) for T and T_h: Delete subscripts 1 and 2 from A and F.

(5) *Rules for* $\mathbf{C}_{\infty v}$ *and* $\mathbf{D}_{\infty h}$:

$$\Sigma^+ \oplus \Sigma^+ = \Sigma^- \oplus \Sigma^- = \Sigma^+, \qquad \Sigma^+ \oplus \Sigma^- = \Sigma^-;$$

$$\Sigma^+ \oplus \Pi = \Sigma^- \oplus \Pi = \Pi,$$

$$\Sigma^+ \oplus \Delta = \Sigma^- \oplus \Delta = \Delta, \quad \text{etc.};$$

$$\Pi \oplus \Pi = \Sigma^+ + \Sigma^- + \Delta,$$

$$\Delta \oplus \Delta = \Sigma^+ + \Sigma^- + \Gamma;$$

$$\Pi \oplus \Delta = \Pi + \Phi.$$

All the multiplications indicated within these rules are commutative. In order to illustrate the above rules consider the product $B_{2u} \oplus E_{1u} \oplus E_{2g}$ in $\mathbf{D}_{6h}$. From

$$u \oplus u \oplus g = u \oplus u = g \tag{18}$$

one concludes that the result belongs to g-representations. From $E_1 \oplus E_2 = B_1 + B_2 + E_1$ the following intermediate result

$$B_{2u} \oplus E_{1u} \oplus E_{2g} = B_{2u} \oplus [B_{1u} + B_{2u} + E_{1u}] \tag{19}$$

is obtained. Because of

$$B_{2u} \oplus B_{1u} = A_{2g}, \qquad B_{2u} \oplus B_{2u} = A_{1g}, \qquad B_{2u} \oplus E_{1u} = E_{2g} \tag{20}$$

one finally obtains

$$B_{2u} \oplus E_{1u} \oplus E_{2g} = A_{1g} + A_{2g} + E_{2g}. \tag{21}$$

8.4 Some Applications of Symmetry Groups

The applications of symmetry groups are too numerous for a complete review. Therefore only a few examples can be given here. Our selection will be limited to topics which have significance in organic chemistry. The various applications of group theory to the theory of the electronic structure of atoms [33, 36, 43, 45] and transition-metal complexes [14, 21, 32, 43, 56, 61] are beyond the scope of this section. Nevertheless it should be mentioned that the atomic quantum numbers l and m can be derived either from the SCHRÖDINGER equation for H, He$^+$ etc., or from the properties of the rotation group $\mathbf{O}(3)$ (see especially Chap. 2 of [36]). Beyond the scope of this book are also the double-groups [14, 33, 36, 42, 49], which are very useful in the discussion of MÖBIUS systems and properties of elementary particles with half-integer quantum numbers of angular momentum, e.g. the electron spin.

8.4.1 Electric Dipole Moment

In general for a given molecule one may define the centers of the positive and the negative electric charge, which depend on the spatial distribution of the nuclei and the electrons, respectively. Due to the electro-neutrality of molecules the absolute values $|Q|$ of the positive and the negative charge are equal. Let R denote the distance vector between the centers of charges. Then the permanent dipole moment μ of the molecule is classically given by $\mu = R\,|Q|$, but within quantum theory it is the expectation value of the electric dipole operator $\mu = er$, where e denotes the charge of an electron and r the radius vector. Thus we have

$$\langle \mu \rangle = \iiint \psi_0^* er\psi_0 \, dx \, dy \, dz \; . \tag{22}$$

Therein ψ_0 denotes the ground state wave function which is assumed to be totally symmetric. According to Eq. (12), the operator μ may be decomposed into its components $\mu_x = ex$, $\mu_y = ey$ and $\mu_z = ez$, which transform like x, y and z, respectively. From Rule 2 it follows that a molecule can have non-zero permanent dipole moment if and only if at least one of the functions x, y, z belongs to the totally symmetric representation of the symmetry group of the molecule.

An inspection of the character tables in Appendix 5 shows that in C_1 μ has arbitrary direction, in C_{1h} μ belongs to the xy-plane, in C_n, C_{nv}, $n \geq 2$ and $C_{\infty v}$ μ coincides with the z-axis and in all other groups μ must be equal to zero.

For example, from the absence of a dipole moment in p-difluorobenzene, one may conclude that the structure of this molecule is D_{2h}. On the other hand the presence of a dipole moment in o- and m-difluorobenzene ($38.5 \cdot 10^{-30}$ Cm and $25.3 \cdot 10^{-30}$ Cm, respectively) indicates that these molecules exhibit C_{2v} or even lower symmetry.

8.4.2 Polarizability

An external electric field E induces an electric dipole $\mu^{(i)}$

$$\mu^{(i)} = \alpha E \tag{23}$$

where α denotes the polarizability tensor, a symmetric square matrix of order 3:

$$\alpha = \begin{pmatrix} \alpha_{xx} & \alpha_{xy} & \alpha_{xz} \\ \alpha_{yx} & \alpha_{yy} & \alpha_{yz} \\ \alpha_{zx} & \alpha_{zy} & \alpha_{zz} \end{pmatrix} \tag{24}$$

$$\alpha_{xy} = \alpha_{yx} \, , \qquad \alpha_{yz} = \alpha_{zy} \, , \qquad \alpha_{zx} = \alpha_{xz} \; . \tag{25}$$

The entries of α transform as the functions indicated as index, i.e. α_{xx} transforms like x^2, α_{xy} like xy, etc.

By a unitary transformation (see Appendices 1 and 3) the matrix α can be diagonalized to α'. Its diagonal elements α'_{xx}, α'_{yy} and α'_{zz} are the main axes of the so-called polarizability ellipsoid. For completely anisotropic molecules, $\alpha'_{xx} \neq \alpha'_{yy} \neq \alpha'_{zz}$. As seen from the character tables in the Appendix 5, this may be realized in the case of rotation groups with a principal axis C_n, $n \leq 2$. For completely isotropic molecules one has $\alpha'_{xx} = \alpha'_{yy} = \alpha'_{zz}$ i.e. the polarizability ellipsoid degenerates to a sphere. This case is realized for molecules belonging to symmetry groups with more than one C_n, $n > 2$. In the case of all other groups, $\alpha'_{xx} = \alpha'_{yy} \neq \alpha'_{zz}$ i.e. the polarizability is represented by an ellipsoid which has prolate or oblate shape, depending on whether $\alpha'_{xx} = \alpha'_{yy} > \alpha'_{zz}$ (e.g. benzene, $C_6H_6 \in \mathbf{D}_{6h}$) or $\alpha'_{xx} = \alpha'_{yy} < \alpha'_{zz}$ (e.g. all molecules of $\mathbf{C}_{\infty v}$ and $\mathbf{D}_{\infty h}$ symmetry), respectively.

The sphape of the polarizability ellipsoid plays a significant role for the refraction indices of polarized light, for the KERR effect and in RAMAN spectroscopy.

8.4.3 Motions of Atomic Nuclei: Translations, Rotations and Vibrations

The symmetry group of a molecule is derived from the equilibrium positions of the nuclei. But even at the zero point of absolute temperature they are not at rest. Thus, one has to attribute three components of the *elongation vectors*, ξ_j, η_j, ζ_j, to each nucleus j, which span a *local cartesian coordinate system* at each nucleus, having its origin in the equilibrium position of the respective nucleus. For convenience ξ_j, η_j and ζ_j are chosen with equal lengths in such a way that they are parallel with the x-, y- and z-axis, respectively, of the molecular coordinate system. The origin of this latter system is in the center of gravity of the molecule. The principal axis of symmetry coincides with the z-axis. In a molecule with n atoms there are in total $3n$ such vector components. They are independent and span a $(3n)$-dimensional vector space, useful for discussing the modes of the motion of the atomic nuclei.

The functions

$$T_x = \Sigma\, \xi_j, \qquad T_y = \Sigma\, \eta_j, \qquad T_z = \Sigma\, \zeta_j \tag{26}$$

describe the translation of the molecule as a whole in the directions of x, y and z, by a unit length. They are the *normal modes of translations*. An arbitrary translation of the molecule is described by a linear combination of T_x, T_y and T_z. It is readily seen that T_x, T_y and T_z transform under the operation of the group like x, y and z, respectively.

There are three other functions derived by considering an infinitesimal rotation of the whole molecule about the x-, y- and z-axis, respectively [66]:

$$R_x = \Sigma\, (z_j\eta_j - y_j\zeta_j), \qquad R_y = \Sigma\, (x_j\zeta_j - z_j\xi_j), \qquad R_z = \Sigma\, (y_j\xi_j - x_j\eta_j) \tag{27}$$

where x_j, y_j and z_j denote the coordinate of the j-th atom in the molecular coordinate system (see Eq. (7.31) and Fig. 7.2f). They are the *normal modes of rotation*. The rotation about an arbitrary axis may be expressed by a linear combination of R_x, R_y and R_z. The transformation properties of R_x, R_y and R_z are assigned in the character tables. As seen from (27), in the case of a collinear molecule (for which $x_j = 0, y_j = 0$ for all j), R_x and R_y have equal absolute values while $R_z = 0$.

Separating the subspaces of the translations and rotations, a subspace of dimension $3n - 6$ (in the case of collinear molecules $3n - 5$) is retained which is the space of the *normal modes of vibration*.

In order to find the normal modes of the motion of all atomic nuclei, the $(3n)$-dimensional space spanned by the elongation vectors is transformed into subspaces corresponding to the irreducible representations of the group, which are then analysed, representation by representation. For illustration, such an analysis is performed here for the water molecule, H_2O.

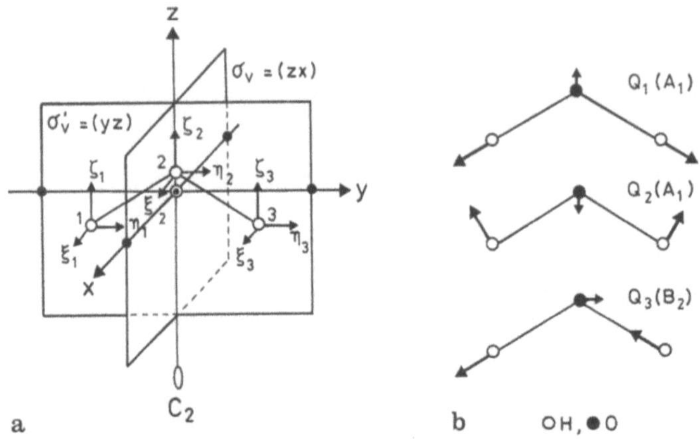

Fig. 8.2. Vibrations in H_2O molecule: **a** location of the atoms, the symmetry elements of C_{2v} and local cartesian coordinate systems; **b** vibrational modes

H_2O belongs to $C_{2v} = \{E, C_2, \sigma_v, \sigma'_v\}$. In Fig. 8.2a the elements of this group and the location of the atoms in $\sigma'_v (= yz$-plane) as well as the local coordinate systems are shown. In an analogous manner to the case of ammonia, Eqs. (7.26) to (7.29), the symmetry coordinates are obtained as

$$A_1: \eta_1 - \eta_3, \qquad \zeta_1 + \zeta_3, \qquad \zeta_2$$
$$A_2: \xi_1 - \xi_3 \tag{28}$$
$$B_1: \xi_1 + \xi_3, \qquad \xi_2$$
$$B_2: \eta_1 + \eta_3, \qquad \eta_2, \qquad \zeta_1 - \zeta_3.$$

The subspace of A_1 is 3-dimensional; since $T_z \in A_1$, there are two vibrations, $Q_1, Q_2 \in A_1$. As seen from Eqs. (28) the single function of A_2 corresponds to R_z (note that in the case of H_2O $y_1 = -y_3$, $y_2 = 0$ and $x_j = 0$ for all j). Hence no vibration belongs to A_2. The same is true for B_1, because $T_x, R_y \in B_1$. Finally, in the 3-dimensional subspace of B_2, in addition to $T_y, R_x \in B_2$ there is also a vibration $Q_3 \in B_2$. For the vibrations the following expressions are obtained:

$$Q_1(A_1) = -k_{11}(\eta_1 - \eta_3) - k_{12}(\zeta_1 + \zeta_3) + k_{13}\zeta_2$$
$$Q_2(A_1) = -k_{21}(\eta_1 - \eta_3) + k_{22}(\zeta_1 + \zeta_3) - k_{23}\zeta_2 \tag{29}$$
$$Q_3(B_2) = -k_{31}(\eta_1 + \eta_3) + k_{32}\eta_2 - k_{33}(\zeta_1 - \zeta_3).$$

The coefficients k_{ij} depend on the mass of the nuclei involved in the vibration, the force constants of the particular symmetry functions and their coupling constants, and they are determined as the solutions of an eigenvalue problem; for details see for instance [66]. In Fig. 8.2b the normal modes of vibrations are depicted; the elongation of O is drawn about four times as long as it really is relative to the elongations of H. The phase of the vibrations agrees with the notation in Eq. (29). Q_1 may be recognized as a nearly pure *symmetric bond stretching mode*, Q_2 as a nearly pure *symmetric bond angle deformation mode*. The B_2-species, Q_3, is called the *asymmetric bond stretching mode*.

Obviously, all bent molecules of A_2B-type belong to C_{2v}; thus Eqs. (28) and (29) are valid for the whole class of this type, but due to the dependence of the coefficients k_{ij} on the molecular parameters (as listed above), the pattern of the A_1-modes varies significantly. So, for example, in sulphur dioxide SO_2, the symmetric bond stretching and bond angle deformation mode are strongly coupled. Therefore in this case Q_1 and Q_2 are mixtures of these internal coordinates [37].

Each normal mode of vibration, Q_k, is associated with a series of *vibrational levels*, labeled by different values of the *vibrational quantum number* v_k, which takes values 0, 1, 2, ... with ascending energy of the corresponding level. Thus, a *vibrational quantum state* is defined by an N-tuple of quantum numbers, $v_1, v_2, ... , v_N$, associated with the respective modes $Q_1, Q_2, ... , Q_N$, where $N = 3n - 6$ and $N = 3n - 5$ for non-linear and collinear molecules, respectively.

At the *ground level* of vibration all quantum numbers are zero, $v_1 = v_2 = ...$ $= v_N = 0$. It can be shown that the ground level always belongs to the totally symmetric representation of the group of the molecule.

The levels where only one quantum number, say v_k, differs from zero are called *fundamental levels* if $v_k = 1$ or *overtone levels* if $v_k \geq 2$. The fundamental level of Q_k, i.e. $v_k = 1$, belongs to the same irreducible representation as Q_k does, say $\Gamma(Q_k)$. The r-th overtone level of Q_k, i.e. $v_k = r$, belongs to the direct product representation obtained from the r-th power of $\Gamma(Q_k)$. In the case when $\Gamma(Q_k)$ is one-dimensional, it belongs to $\Gamma(Q_k)$ if r is odd, but to the totally symmetric representation if r is even.

In *combination levels* more than one quantum number differs from zero; they belong to the representation which results from the direct product of representations to which the fundamentals and overtones belong.

When vibrational modes are treated in the so-called harmonic approximation [66] then transitions between the various levels are allowed only for a change of a single quantum number by unity *(harmonic oscillator selection rule)*. This rule is sometimes violated due to the anharmonicity of the potential energy function for the motion of the nuclei [37] (see the subsequent paragraph).

8.4.4 Transition Probabilities for the Absorption of Light

Light has a dualistic nature: in some experiments it behaves as an electromagnetic wave of frequency v; in other experiments it exhibits properties of corpuscles (photons) of energy hv and momentum hv/c, where h denotes the PLANCK constant and c the velocity of light.

In quantum chemistry the interaction between molecules and light is sufficiently well described by means of time-dependent perturbation theory (see for instance Chap. 8 of [20]). It is shown that a transition from an initial quantum state ψ_m of energy E_m into a final one ψ_n of energy E_n takes place by absorbing a photon of the appropriate energy $h\nu = E_n - E_m$. The probability of such a transition is proportional to the square[1] $|R_{mn}|^2$. There are various contributions to R_{mn} which are classified as electric or magnetic dipoles, quadrupoles etc. They differ by several orders of magnitude[2]. Thus R_{mn} may be sufficiently well approximated by the electric dipole contribution only, the so-called *electric transition dipole*, defined as

$$R_{mn} = \int \psi_n^* R \psi_m \, d\tau \tag{32}$$

where ψ_m and ψ_n denote state functions, R the vector sum of all electronic position vectors, $R = x + y + z$, and $d\tau$ the volume element built up from the coordinates of all electrons. Replacing R in Eq. (32) by its vector components, the integrals X_{mn}, Y_{mn} and Z_{mn} are defined in an analogous manner, which satisfy the equation

$$|R_{mn}|^2 = |X_{mn}|^2 + |Y_{mn}|^2 + |Z_{mn}|^2 \,. \tag{33}$$

From Rule 2 one concludes that the transition $\psi_m \to \psi_n$ has non-zero probability only if at least one of the integrands $\psi_n^* x \psi_m$, $\psi_n^* y \psi_m$ or $\psi_n^* z \psi_m$ belongs to the totally symmetric representation. If this is the case, then such a transition is said to be *symmetry-allowed*. For example, according to (21), this would be the case for the $B_{2u} \to E_{2g}$ transition in $\mathbf{D}_{6h}$, where $x, y \in E_{1g}$. Since for that transition Z_{mn} is identically zero, $Z_{mn} = 0$, one says that the transition is x, y-*polarized*.

If for a particular transition $X_{mn} = Y_{mn} = Z_{mn} = 0$ and, hence, $R_{mn} = 0$, one says that the transition is *symmetry-forbidden*. The so-called $n \to \pi^*$ transitions are symmetry-forbidden with respect to the electric transition dipole R_{mn}. They are, however, rendered by the *magnetic transition dipole* M_{mn} defined by

$$M_{mn} = \int \psi_n^* R \times P \psi_m d\tau \tag{34}$$

where P denotes the momentum and, thus, the vector product $R \times P$ is the angular momentum of the electrons. Due to the difference in the respective orders of magnitude of R_{mn} and M_{mn}, $n \to \pi^*$ transitions have low intensities.

In the discussion above ψ_m and ψ_n have been understood to be electronic state functions. The *symmetry selection rule* given above applies also when ψ_m and ψ_n represent *vibrational quantum states* within a given electronic configuration. Then, however, the symmetry selection rule is superimposed on the harmonic oscillator selection rule mentioned in the foregoing paragraph. Consequently, the transitions from the totally symmetric ground level to any fundamental level have non-zero probability only for those normal coordinates Q_k which belong to an irreducible representation to which at least one of the coordinates x, y and z also belongs. These

---

[1] In general R_{mn} can be a complex number. In that case $|R_{mn}|^2 = R_{mn}^* R_{mn}$.
[2] The contributions of magnetic dipoles and electric quadrupoles are by a factor of 10^{-5} and 10^{-7}, respectively, smaller than the electric dipole contribution to R_{mn}.

normal coordinates are called *active modes*, the other ones are *inactive*. As seen from the analysis presented in the foregoing paragraph and from the character table of C_{2v}, all the possible vibrational transitions in H_2O are symmetry-allowed. Thus the IR-spectrum of H_2O should be governed by the harmonic oscillator selection rule, $\Delta v_k = \pm 1$. But, as already mentioned, this rule is violated due to the anharmonicity of the potential [37]. Therefore the IR bands of the H_2O vapour also contain transitions ascribed as $(000) \rightarrow (020)$, $(000) \rightarrow (011)$ and $(000) \rightarrow (021)$ at wave numbers $\bar{v} = 3151$, 5332 and 6874 cm^{-1}, respectively. The notation (v_1, v_2, v_3) is in accordance with Q_1, Q_2 and Q_3 as depicted in Fig. 8.2b.

In C_{2v} IR-inactive modes are those which belong to A_2. Such a vibrational mode exists for difluoromethane, CH_2F_2, as the reader may derive as an exercise. The A_2-mode of CH_2F_2 corresponds to the twisting of the planes of the CH_2 and CF_2 groups, thus varying the dihedral angle θ between these two planes. The equilibrium value for θ is $\pi/2$ (see Fig. 8.3).

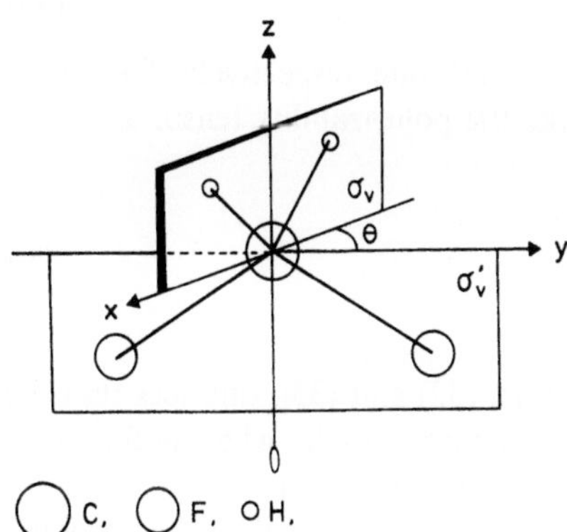

Fig. 8.3. The CH_2F_2 molecule in C_{2v}

The symmetry selection rule holds also for transitions between rotational as well as rotation-vibrational levels. We shall not discuss it because the pure rotational and the rotation-vibrational spectra are of nearly negligible use in organic chemistry. Note, however, that the rotational spectra obtained by microwave absorption are in general a powerful tool in determining internuclear distances and angles.

8.4.5 Transition Probabilities in Raman Spectra

As discussed in the foregoing paragraph, monochromatic light of frequency v can be absorbed by a molecule in a state of energy E_0 only if it has another state of energy E_1, such that $E_1 - E_0 = hv$. If this is not the case, then the irradiated light is scattered.

Suppose that the energy of the irradiating light is rather high compared with the quantum hv_k of any vibrational excitation of the molecule, but much lower than the electronic energy, $hv_k \ll hv \ll E_1 - E_0$. Then its energy is insufficient for absorption. An analysis of the scattered light shows that in addition to the initial frequency (RAYLEIGH scattering), some frequencies $v - v_k$ and $v + v_k$ are also present (RAMAN scattering), with intensities due to the population of the ground and the fundamental level, respectively (see Fig. 8.4).

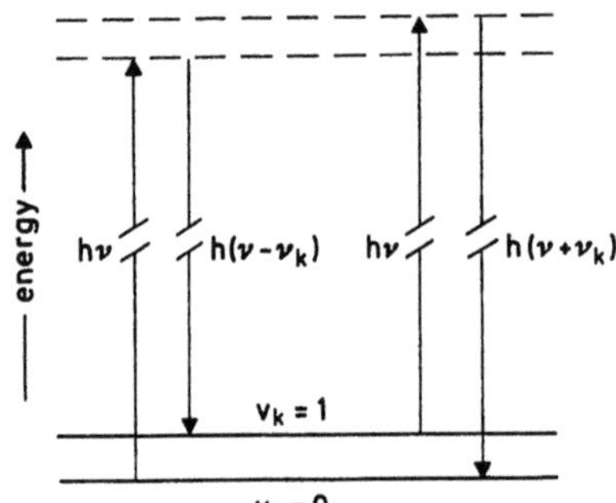

Fig. 8.4. RAMAN scattering: the interaction of a molecule with an energy quantum $h\nu$ insufficient for absorption

From time-dependent perturbation theory one obtains [20] that the probability of both RAMAN transitions is proportional to the square of the integral

$$\int \psi_n^* R \cdot R \psi_m \, d\tau \tag{35}$$

where ψ_m and ψ_n denote the initial and final vibrational state, respectively. The tensor product $R \cdot R$ of the radius vector transforms like the polarizability tensor as

$$R \cdot R = \begin{pmatrix} x^2 & xy & xz \\ yx & y^2 & yz \\ zx & zy & z^2 \end{pmatrix}. \tag{36}$$

Comparing the IR and RAMAN transitions, Eqs. (32) and (35), one sees that the former are made feasible by a change of the permanent dipole, whereas the latter by a change of the induced dipole due to the vibration considered.

As seen from (35), a RAMAN transition is symmetry-allowed if the direct product $\Gamma(\psi_m) \oplus \Gamma(\xi_i\xi_j) \oplus \Gamma(\psi_n)$ is either totally symmetric or contains the totally symmetric representation; $\xi_i\xi_j$ denotes any of the bilinear forms x^2, xy etc. The character table of C_{2v} shows that each irreducible representation of this group has at least one of these species. Thus, the RAMAN spectrum of a molecule belonging to C_{2v} contains the fundamental tones of all vibrations.

From the character table of $D_{\infty h}$ is seen that the totally symmetric representation Σ_g^+ has two such species, $x^2 + y^2$ and z^2. Hence, the fundamental transition of the symmetric stretching mode of CO_2 which belongs to Σ_g^+ is present in the RAMAN spectrum, in contrast to the IR spectrum where it is absent (see the foregoing paragraph). This may serve as an example for the so-called IR-RAMAN alternative rule.

IR-RAMAN **Alternative Rule:** In a centro-symmetric molecule (which has a center of inversion) the linear functions x, y and z belong to u-representations, their bilinear forms x^2, y^2, z^2, xy, yz and zx to g-representations. Thus in view of the transformation properties of vibrational states, the IR and RAMAN spectra of such a molecule can obtain fundamental absorption of u- and g-modes, respectively. This means that the fundamentals present in the IR absorption spectrum are absent in the RAMAN spectrum and vice versa.

In the case of benzene, C_6H_6, belonging to D_{6h}, one meets the following situation. From the 30 vibrational modes 15 belong to u- and 15 to g-representations. Among these modes, 7 are active in the IR spectrum and 12 in the RAMAN spectrum. Among the 12 irreducible representations of D_{6h} only 5 are active in the IR or in the RAMAN spectrum.

8.4.6 Group Theory and Quantum Chemistry

Some results of the numerous applications of group theory to quantum chemistry [11, 20, 36] have already been used in the previous parts of this section. Here a few additional features will be outlined.

The central problem of quantum chemistry is to find the eigenvalues E_k and the eigenfunctions ψ_k of the Hamiltonian operator $\mathscr{H}$ of the system considered. Most commonly this problem is represented by the (non-relativistic) time-independent SCHROEDINGER equation

$$\mathscr{H}\psi_k = E_k\psi_k \,. \tag{37}$$

It can be shown that the following general results hold.

Rule 6: The Hamiltonian $\mathscr{H}$ of a given molecule always belongs to the totally symmetric representation of the group of the molecule considered.

Rule 7 (Wiegner's theorem): The molecular Hamiltonian $\mathscr{H}$ commutes with each element R of the group G of the molecule.

Let $R\psi_k$ denote the function obtained from ψ_k under the symmetry operation R of G. If ψ_k satisfies Eq. (37), then according to rule 7,

$$\mathscr{H}(R\psi_k) = R(\mathscr{H}\psi_k) = R(E_k\psi_k) = E_k(R\psi_k) \,. \tag{38}$$

From the invariance of the eigenvalues under the operation of the group one reaches the following important conclusion.

Rule 8: The eigenfunctions of a molecular Hamiltonian transform according to particular irreducible representations of the symmetry group of the molecule.

Since Eq. (37) can be solved exactly for only a few cases, numerous approximate methods have been developed. Let φ_1, φ_2, ... be a set of functions which are symmetry adapted, i.e. they belong to particular representations, $\varphi_j \in \Gamma(\varphi_j)$ of the symmetry group of the molecule. For example, the functions φ_j may be atomic orbitals, linear combinations of them, etc., but by no means eigenfunctions of the molecular Hamiltonian. Further let these functions depend on a given molecular parameter q, e.g. the coordinates of the equilibrium positions of the nuclei.

The molecular Hamiltonian $\mathscr{H}$ can be then represented within this basis by means of a square matrix H whose entries are

$$H_{jk} = \int \varphi_k^* \mathscr{H} \varphi_j \, d\tau \,. \tag{39}$$

Then the matrix H has diagonal block form and each block corresponds to a particular irreducible representation of the group. This is because from Rules 2 and 6 it follows that H_{jk} is identically zero if φ_j and φ_k belong to different representations. The diagonal elements H_{jj} represent energies associated with the functions φ_j while the off-diagonal elements $H_{jk}, j \neq k$, indicate the energetic effect of the interaction of the functions φ_j and φ_k. Since the functions φ_j are assumed to depend on the molecular parameter q, all the matrix elements of H must be functions of this parameter: $H_{jk} = H_{jk}(q)$.

For a distinct value of q, say q', it may happen that two diagonal elements H_{jj} and H_{kk} take the same value. Then the functions $H_{jj}(q)$ and $H_{kk}(q)$ intersect each other at $q = q'$ as shown schematically in Fig. 8.5a.

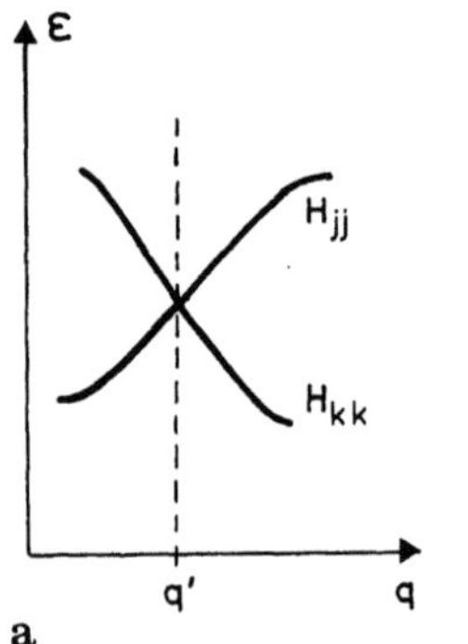

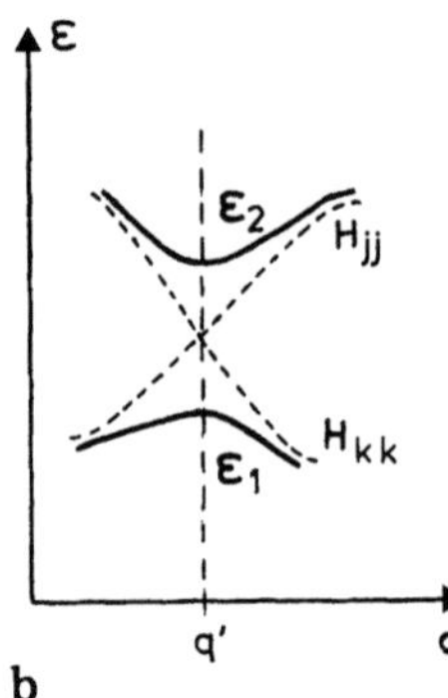

Fig. 8.5. Illustration of the non-crossing rule: **a** two crossing levels; **b** removed crossing

In this case an eigenvalue problem must be solved for the functions φ_j and φ_k at $q = q'$ (see especially Sect. C of Chap. 16 of [20]), which results in

$$\varepsilon_{1,2} = [H_{jj} + H_{kk} \pm \sqrt{(H_{jj} - H_{kk})^2 + 4H_{jk}^2}]/2 . \tag{40}$$

If φ_j and φ_k belong to the same irreducible representation, $\Gamma(\varphi_j) \equiv \Gamma(\varphi_k)$, then H_{jk} is not necessarily zero and the crossing is removed (Fig. 8.5b). This result is the essence of the non-crossing rule.

Rule 9 (the non-crossing rule): Two levels belonging to the same irreducible representation never intersect.

If φ_j and φ_k belong to different representations, $\Gamma(\varphi_j) \neq \Gamma(\varphi_k)$, then according to Rules 2 and 6, H_{jk} is identically zero and consequently, Eq. (40) results in $\varepsilon_{1,2} \in \{H_{jj}, H_{kk}\}$, i.e. two levels cannot interact. Hence the crossing remains as depicted in Fig. 8.5a.

Although Rule 9 is strictly valid only if the crossing levels depend on a single parameter, e.g. the distance between the nuclei in a diatomic molecule, its applications are much wider.

8.4.7 Orbital and State Correlations

Some groups may be generated as products of smaller groups according to (7.45). In this procedure the irreducible representations of one group are split by those of the other group. This has been briefly discussed for $\mathbf{C}_{nv} = \mathbf{C}_n \otimes \{E, \sigma_v\}$ and $\mathbf{C}_{nh} = \mathbf{C}_n \oplus \{E, \sigma_h\}$ in Sect. 8.2. Here the interest is focused on exactly the opposite: What happens to the irreducible representations if the symmetry of the molecule is lowered? The remaining symmetry group is a proper subgroup of the original group.

The number of different subgroups which can be extracted from the group considered is equal to the number of ways in which the original symmetry can be lowered. Let us take C_{2v} as an example and let us lower the C_{2v} symmetry once to C_2 and subsequently to C_{1h}:

C_2	E	C_2
A	1	1
B	1	-1

C_{2v}	E	C_2	σ_v	σ_v'
A_1	1	1	1	1
A_2	1	1	-1	-1
B_1	1	-1	1	-1
B_2	1	-1	-1	1

C_{1h}	E	σ
A'	1	1
A''	1	-1

In the case of the symmetry lowering $C_{2v} \to C_2$, the twofold axis is retained. In $A \in C_2$ one has $\chi(C_2) = 1$; the same characters one finds in C_{2v} for A_1, $A_2 \in C_{2v}$. Thus, the functions belonging in C_{2v} either to A_1 or A_2 will belong to A in C_2. One says concisely that A_1 and A_2 of C_{2v} *correlate with* A of C_2. From $\chi(C_2) = -1$ one similarly concludes that B_1, $B_2 \in C_{2v}$ correlate with $B \in C_2$.

In the case of the symmetry lowering $C_{2v} \to C_{1h}$ first of all it must be decided which of the two reflection planes of C_{2v}, σ_v or σ_v', is retained. Let us assume that σ_v is retained. Then from the values of $\chi(\sigma_v)$ in C_{2v} and C_{1h} it follows that A_1 and B_1 of C_{2v} correlate with A' of C_{1h}. Similarly, A_2, $B_2 \in C_{2v}$ correlate with $A'' \in C_{1h}$. If the symmetry species correlated are MO's, one speaks about *orbital correlation*, whereas if they are states, one uses the term *state correlation*.

As an illustration the ring closure of cis-butadiene (I) to cyclobutene (II) and the inverse reaction of ring opening, II $\to$ I will be considered:

$$\text{I} \quad\rightleftharpoons\quad \text{II}$$

Both molecules have C_{2v} symmetry. In the reaction I $\to$ II two π bonds are broken and a new π bond as well as an additional σ bond are formed. In the reaction II $\to$ I the opposite happens. The reaction requires a rotation of the CH_2 groups about the respective CC bonds since they are coplanar in I but perpendicular to the plane of the carbon atoms in II. This rotation lowers the original C_{2v} symmetry. If both planes rotate with equal velocities in the same direction (*conrotatory mode*), then the C_2 axis is conserved and the symmetry is lowered from C_{2v} to C_2. However, if the directions of the rotations are opposite (*disrotatory mode*), then the σ_v plane is conserved and the symmetry lowering occurs from C_{2v} to C_{1h}.

In Fig. 8.6 those MO's of I and II are schematically depicted which are subject to significant changes in the course of the reaction. They are readily classified as π, π^*, σ and σ^* MO's. Their respective irreducible representations are indicated. As a consequence of the different correlation of the C_{2v} species with those of C_2 and C_{1h}, the MO's of cis-butadiene (I) and cyclobutadiene (II) correlate in a different manner for the con- and the disrotatory mode. The orbital correlation shown in Fig. 8.6 represents one of the numerous applications of WOODWARD-HOFFMANN rules to organic reactions [2, 68].

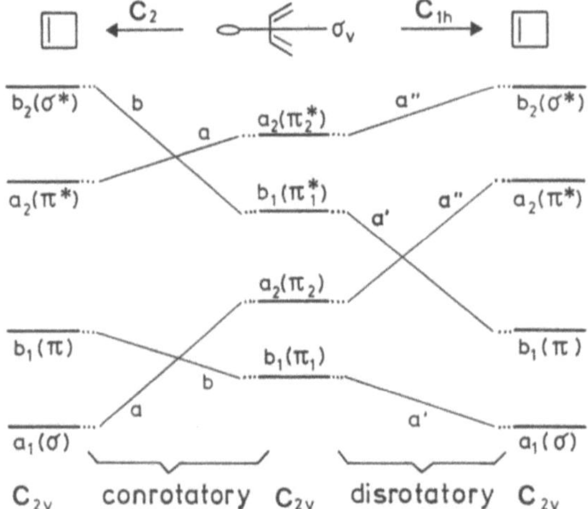

Fig. 8.6. Orbital correlation for the reaction I ⇌ II

The conrotatory and the disrotatory modes are competing reaction paths. In this competition that path is favoured which needs a lower energy of activation. Qualitative conclusions about the relative magnitude of the energy of activation may be drawn from the state correlation diagram (Fig. 8.7). Therein the states involved in the thermal and in the photoreaction I → II and II → I are denoted by ψ_i^J, $i = 1, 2, 3, 4$; $J = $ I, II. For the thermal reaction the reactant is assumed to be in

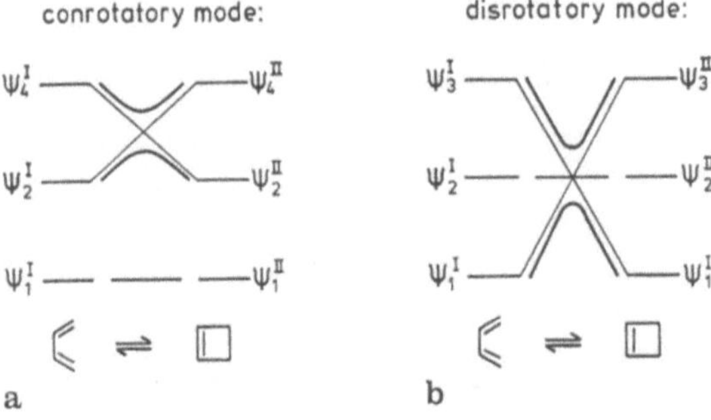

Fig. 8.7. State correlation for the reaction I → II. The conrotatory pathway is preferred for thermal, the disrotatory pathway for photoreaction

its ground state ψ_1^J, whereas in the photochemical reaction it is in its first excited state ψ_2^J. The other states involved are the lowest doubly excited state ψ_3^J and another singly excited state ψ_4^J. From an inspection of Fig. 8.6 and the character table of C_{2v} one finds

$$\psi_1^{\mathrm{I}} = \pi_1^2\pi_2^2 \in \boldsymbol{A}_1\,, \qquad\qquad \psi_1^{\mathrm{II}} = \sigma^2\pi^2 \in \boldsymbol{A}_1\,,$$

$$\psi_2^{\mathrm{I}} = \pi_1^2\pi_2\pi_1^* \in \boldsymbol{B}_2\,, \qquad\qquad \psi_2^{\mathrm{II}} = \sigma^2\pi\pi^* \in \boldsymbol{B}_2\,,$$

$$\psi_3^{\mathrm{I}} = \pi_1^2\pi_1^{*2} \in \boldsymbol{A}_1\,, \qquad\qquad \psi_3^{\mathrm{II}} = \sigma^2\pi^{*2} \in \boldsymbol{A}_1\,, \tag{41}$$

$$\psi_4^{\mathrm{I}} = \pi_1\pi_2^2\pi_2^* \in \boldsymbol{B}_2\,; \qquad\qquad \psi_4^{\mathrm{II}} = \sigma\pi^2\sigma^* \in \boldsymbol{B}_2\,.$$

The Figures 8.7a and 8.7b show schematically the state correlations in the case of the conrotatory and disrotatory mode, respectively. In the conrotatory mode the levels of the photoreaction, $\psi_2^{\mathrm{I}} \to \psi_4^{\mathrm{II}}$ and $\psi_2^{\mathrm{II}} \to \psi_4^{\mathrm{I}}$, in the disrotatory mode those of the thermal reaction $\psi_1^{\mathrm{I}} \to \psi_3^{\mathrm{II}}$ and $\psi_1^{\mathrm{II}} \to \psi_3^{\mathrm{I}}$, cross each other. Due to Rule 9 they must not cross. Since an avoided crossing is usually associated with a strong increase of energy, from Fig. 8.7 one may conclude that the thermal reaction I → II follows the conrotatory reaction path while the photoreaction takes place in the disrotatory mode. These conclusions are experimentally confirmed [141].

Although the WOODWARD-HOFFMANN rules are usually discussed in terms of symmetry as above, one should realize that they are essentially topological. Hence, they work even if the molecules considered have no higher symmetry than $\boldsymbol{C}_1$.

In order to save space, no correlation tables for descent in symmetry are given here; they can be found in [58].

Chapter 9

Automorphism Groups

9.1 Automorphism of a Graph

The notion of graph automorphism has already been introduced in Section 4.1. An automorphism may be understood as a bijective (that is one-to-one) mapping of the vertex set $\mathscr{V}(G)$ of the graph onto itself which preserves the edge relation $\mathscr{E}(G)$ of the graph. Evidently, only those vertices can be mapped onto each other which are equivalent, i.e. they are indistinguishable apart from their labels. A subset of $\mathscr{V}(G)$ formed by all mutually equivalent vertices is called an *orbit* of the graph vertices.

As seen by inspection the vertices of the graph G_1 form four orbits, namely:

$$\{1\}, \qquad \{2\}, \qquad \{3, 4\}, \qquad \{5, 6, 7\}. \tag{1}$$

Obviously, the vertices 1 and 2 can be mapped automorphically only onto themselves. For the vertices of $\{3, 4\}$ and $\{5, 6, 7\}$ there are 2 and 6 feasible bijective mappings, respectively. Since each automorphism of G_1 can be regarded as a particular combination of just one bijective mapping of equivalent vertices one has $1 \cdot 1 \cdot 2 \cdot 6 = 12$ different automorphisms for G_1. In the next section it is shown that the omnium of the automorphisms of a graph forms a group, the so-called automorphism group of the graph.

As shown in Sect. 4.1 an automorphism is unequivocally expressed by an appropriate *permutation*: in the upper line the origins, in the lower line the images of the corresponding automorphic mapping are denoted.

The combination of two permutations is explained as follows. (i) The permutation at the right-hand side is assumed to be applied first; its lower line assigns the images of the originals. (ii) The permutation at the left-hand side, now applied, operates upon these images. (iii) The result is represented by another permutation which correlates the originals with the final images. In this way one has for example:

$$\begin{pmatrix} 1 & 2 & 3 & 4 & 5 & 6 & 7 \\ 1 & 2 & 4 & 3 & 6 & 7 & 5 \end{pmatrix} \begin{pmatrix} 1 & 2 & 3 & 4 & 5 & 6 & 7 \\ 1 & 2 & 4 & 3 & 5 & 7 & 6 \end{pmatrix} = \begin{pmatrix} 1 & 2 & 3 & 4 & 5 & 6 & 7 \\ 1 & 2 & 3 & 4 & 6 & 5 & 7 \end{pmatrix}. \tag{2}$$

9.2 The Automorphism Group A(G_1)

The orbits of equivalent vertices of G_1 are given by (1); they have the cardinalities 1, 1, 2 and 3, respectively. The automorphisms of G_1 already discussed above are represented by the following permutations which are abbreviated by letters for convenience:

$$E:\begin{pmatrix}1&2&3&4&5&6&7\\1&2&3&4&5&6&7\end{pmatrix},\quad A:\begin{pmatrix}1&2&3&4&5&6&7\\1&2&3&4&6&7&5\end{pmatrix},\quad B:\begin{pmatrix}1&2&3&4&5&6&7\\1&2&3&4&7&5&6\end{pmatrix},$$

$$C:\begin{pmatrix}1&2&3&4&5&6&7\\1&2&3&4&5&7&6\end{pmatrix},\quad D:\begin{pmatrix}1&2&3&4&5&6&7\\1&2&3&4&7&6&5\end{pmatrix},\quad F:\begin{pmatrix}1&2&3&4&5&6&7\\1&2&3&4&6&5&7\end{pmatrix},$$

$$H:\begin{pmatrix}1&2&3&4&5&6&7\\1&2&4&3&5&6&7\end{pmatrix},\quad J:\begin{pmatrix}1&2&3&4&5&6&7\\1&2&4&3&6&7&5\end{pmatrix},\quad K:\begin{pmatrix}1&2&3&4&5&6&7\\1&2&4&3&7&5&6\end{pmatrix},$$

$$L:\begin{pmatrix}1&2&3&4&5&6&7\\1&2&4&3&5&7&6\end{pmatrix},\quad M:\begin{pmatrix}1&2&3&4&5&6&7\\1&2&4&3&7&6&5\end{pmatrix},\quad N:\begin{pmatrix}1&2&3&4&5&6&7\\1&2&4&3&6&5&7\end{pmatrix}.$$

$$(3)$$

By means of Eq. (2) the multiplication table of these permutations is obtained as follows:

	E	A	B	C	D	F	H	J	K	L	M	N
E	E	A	B	C	D	F	H	J	K	L	M	N
A	A	B	E	F	C	D	J	K	H	N	L	M
B	B	E	A	D	F	C	K	H	J	M	N	L
C	C	D	F	E	A	B	L	M	N	H	J	K
D	D	F	C	B	E	A	M	N	L	K	H	J
F	F	C	D	A	B	E	N	L	M	J	K	H
H	H	J	K	L	M	N	E	A	B	C	D	F
J	J	K	H	N	L	M	A	B	E	F	C	D
K	K	H	J	M	N	L	B	E	A	D	F	C
L	L	M	N	H	J	K	C	D	F	E	A	B
M	M	N	L	K	H	J	D	F	C	B	E	A
N	N	L	M	J	K	H	F	C	D	A	B	E

$$(4)$$

From the multiplication table one may conclude that the permutations given in (3) indeed form a group under the law of multiplication exemplified by Eq. (2). The axioms [G 1] to [G 4] are satisfied; the permutation E represents the identity element of the group. This group is called the *automorphism group of the graph* G_1, denoted by A(G_1). Its *order* is $h(A(G_1)) = 12$.

Further, from the multiplication table the *inverse elements* may be obtained as

X	E	A	B	C	D	F	H	J	K	L	M	N
X^{-1}	E	B	A	C	D	F	H	K	J	L	M	N

$$(5)$$

Finally, by means of Definition 4 from Chap. 7 the following classes are obtained:

$$\{E\}, \{A, B\}, \{C, D, F\}, \{H\}, \{J, K\}, \{L, M, N\} . \tag{6}$$

Since there are 6 classes, according to Rule 1 from Chap. 7 there must be 6 irreducible representations. Let s_j denote the number of irreducible representations of dimension j, then we have

$$6 = s_1 + s_2 + s_3 + \dots \tag{7}$$

and according to Eq. (7.21)

$$12 = s_1 + 4s_2 + 9s_3 + \dots . \tag{8}$$

From these two equations it follows that $s_1 = 4$, $s_2 = 2$ and $s_j = 0$ for $j \geq 3$. Hence, the automorphism group $A(G_1)$ has four 1-dimensional and two 2-dimensional irreducible representations.

It can be easily shown that $A(G_1)$ may be generated by means of the permutations A, C and H according to

$$\{E, C\} \otimes \{E, A, B\} = \{E, A, B, C, D, F\} \tag{9}$$

$$\{E, H\} \oplus \{E, A, B, C, D, F\} = A(G_1) . \tag{10}$$

Equation (10) can be used for the construction of the character table of $A(G_1)$. Having in mind the character table (7.25) of the group $\{E, A, B, C, D, F\}$, we immediately arrive at (11).

	$\{E\}$	$\{A, B\}$	$\{C, D. F\}$	$\{H\}$	$\{J, K\}$	$\{L, M, N\}$
Γ_1	1	1	1	1	1	1
Γ_1'	1	1	1	-1	-1	-1
Γ_2	1	1	-1	1	1	-1
Γ_2'	1	1	-1	-1	-1	1
Γ_3	2	-1	0	2	-1	0
Γ_3'	2	-1	0	-2	1	0

$$(11)$$

The procedure outlined above shows that all the fundamental laws of group theory, as exemplified in the Chaps. 7 and 8 for spatial symmetry groups, are also applicable to groups formed by permutations. There is only the idea of the *degree* of permutation groups which is undefined in space groups and, hence, must be added here [164]:

Definition 1: The degree of a permutation group is the number of objects upon which the permutations of the group operate.

The degree of the permutation group $\mathbf{G}$ will be denoted by $g(\mathbf{G})$. Obviously, the degree of an automorphism group of a graph is equal to the number of vertices of this graph.

9.3 Cycle Structure of Permutations

Consider the permutation J given in (3). J may be understood as the result of the action of four permutations, operating upon disjoint subsets, namely

$$J: \begin{pmatrix} 1 & 2 & 3 & 4 & 5 & 6 & 7 \\ 1 & 2 & 4 & 3 & 6 & 7 & 5 \end{pmatrix} = \begin{pmatrix} 1 \\ 1 \end{pmatrix} \begin{pmatrix} 2 \\ 2 \end{pmatrix} \begin{pmatrix} 3 & 4 \\ 4 & 3 \end{pmatrix} \begin{pmatrix} 5 & 6 & 7 \\ 6 & 7 & 5 \end{pmatrix}. \tag{12}$$

In (12) the cycle structure of the permutation J comes forward. The labels 5, 6 and 7 are permuted in a three-membered cycle, the labels 3 and 4 in a two-membered cycle and, finally, the labels 1 and 2 in one-membered cycles each. The cycle structure of the permutation J, denoted as $cs(J)$ may be expressed as

$$cs(J) = [1]^2 [2][3] . \tag{13}$$

It is easy to see that if a given permutation P has the cycle structure

$$cs(P) = [1]^\alpha [2]^\beta [3]^\gamma \dots , \tag{14}$$

then the degree of the permutation is given by

$$g(P) = \alpha + 2\beta + 3\gamma + \dots . \tag{15}$$

The cycle structures of the permutations forming the automorphism group $\mathbf{A}(G_1)$ are

$$cs(E) = [1]^7 , \qquad cs(A) = cs(B) = [1]^4 [3] ,$$

$$cs(C) = cs(D) = cs(F) = [1]^5 [2] , \qquad cs(H) = [1]^5 [2] , \tag{16}$$

$$cs(J) = cs(K) = [1]^2 [2][3] , \qquad cs(L) = cs(M) = cs(N) = [1]^3 [2]^2 .$$

The following rule holds.

Rule 1: The permutations belonging to the same class have equal cycle structure.

The opposite is not true as exhibited by $\{H\}$ and $\{C, D, F\}$ in eqs. (16). The non-equivalence of the classes $\{H\}$ and $\{C, D, F\}$ is due to the nonequivalence of the objects transposed by these permutations, $\{3, 4\}$ and $\{5, 6, 7\}$, respectively.

As mentioned in Appendix 2 each permutation may be considered as a result of a number of transpositions. It can be shown that the cycle $[m]$ needs $m - 1$ *transpositions*. Hence, its parity is given by $(-1)^{m-1}$. From this, by inspection of Eqs. (16) one readily finds that the permutations E, A, B, L, M and N are even while C, D, F, H, J and K are odd. Since permutations belonging to the same class have the same cycle structure, they also must have the same parity.

Having in mind that the parity of an identical permutation E is, regardless of its degree, always $(-1)^0 = +1$, from the group axiom [G 2] one easily concludes that no automorphism group can consist solely of odd permutations. On the other hand, permutation groups consisting only of even permutations do exist, as illustrated by the subgroups $\{E, A, B\}$ and $\{E, A, B, L, M, N\}$ of $\mathbf{A}(G_1)$. It can be shown that the following statement holds.

Rule 2: An automorphism group of order h consists either of h even or of $h/2$ even and $h/2$ odd permutations. If h is odd, then all permutations forming the group are even.

The notation on the right-hand side of Eq. (12) is often used to present permutations in a single line. The indication of the objects which are permuted is unequivocal only in the case of the cycle structure $[1]^{\alpha}[2]^{\beta}$. For example, $L = (1)(2)(34)(5)(67)$ and $M = (1)(2)(34)(57)(6)$. If the permutation contains n-cycles, $n > 2$, then a whole set of permutations is indicated by the same symbol. Thus $(1)(2)(34)(567)$ stands for both the permutations J and K.

9.4 Isomorphism of Graphs and of Automorphism Groups

Let G_a and G_b be two isomorphic graphs, $G_a \simeq G_b$. Then the automorphism groups of these graphs must be isomorphic, $A(G_a) \simeq A(G_b)$, because each particular automorphism of G_a is in a one-to-one correspondence with a distinct automorphism of G_b. The reverse of this statement is not true. Figure 9.1 shows examples of non-isomorphic graphs with isomorphic automorphism groups. The reason for the isomorphism of the respective automorphism groups is the fact that all the graphs depicted in Fig. 9.1 have the same number of vertices partitioned into the same number of equivalence classes ($=$ orbits) of pairwise identical cardinalities.

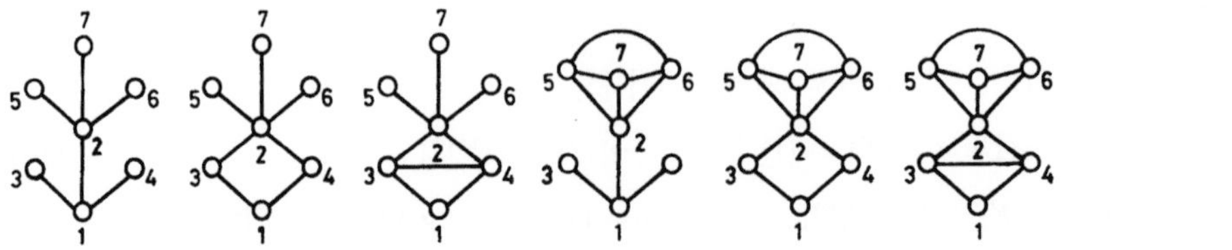

Fig. 9.1. Six non-isomorphic graphs with 7 vertices. The first graph from the left has been already used and is denoted by G_1. The automorphism groups of all graphs shown are isomorphic to $\mathbf{A}(G_1)$

It is worthwhile to note that the automorphism group of a graph depends on the classes of equivalent vertices and their interrelations. In order to illustrate the role of the interrelations between two subsets of equivalent vertices, we compare the automorphisms of the graphs G_2 and G_3. These two graphs differ only in two edges, (3,6) and (4,7), which are present in G_3 but not in G_2. Both graphs have the same orbits, $\{1\}$, $\{2\}$, $\{3, 4\}$, $\{5\}$ and $\{6, 7\}$. Obviously, the automorphism group of G_2 is formed by the permutations

$$G_2 \qquad\qquad\qquad G_3$$

$$E : \begin{pmatrix} 1 & 2 & 3 & 4 & 5 & 6 & 7 \\ 1 & 2 & 3 & 4 & 5 & 6 & 7 \end{pmatrix}; \qquad P : \begin{pmatrix} 1 & 2 & 3 & 4 & 5 & 6 & 7 \\ 1 & 2 & 4 & 3 & 5 & 6 & 7 \end{pmatrix},$$

$$Q : \begin{pmatrix} 1 & 2 & 3 & 4 & 5 & 6 & 7 \\ 1 & 2 & 3 & 4 & 5 & 7 & 6 \end{pmatrix}, \qquad R : \begin{pmatrix} 1 & 2 & 3 & 4 & 5 & 6 & 7 \\ 1 & 2 & 4 & 3 & 5 & 7 & 6 \end{pmatrix}. \tag{17}$$

Thus $A(G_2) = \{E, P, Q, R\}$. Due to the presence of the edges $(3, 6)$ and $(4, 7)$ in G_3, the permutations P and Q are not feasible in the case of G_3. Therefore, $A(G_3) = \{E, R\}$. Note that the equivalent vertices of $\{3, 4\}$ and $\{6, 7\}$ may be automorphically mapped in G_2 independently of each other. Contrary to that, the interrelation between these two orbits, caused in G_3 by the presence of the additional edges, requires a simultaneous well-tuned mapping of the vertices of these two orbits.

9.5 Notation of some Permutation Groups

We list here briefly the most common permutation groups of degree n [36, 38, 41, 46].

$\mathbf{S}_n$　The *symmetric group* of order $h = n!$ consists of all possible permutations of n objects.

$\mathbf{A}_n$　The *alternant group* of order $h = n!/2$ consists of all possible even permutations of n objects.

$\mathbf{D}_n$　The *dihedral group* of order $h = 2n$ is generated by the n-cycle $(1\ 2\ 3\ ...\ n)$ and $(1\ n)\,(2\ n-1)\,(3\ n-2)\ ...$.

$\mathbf{C}_n$　The *cyclic group* of order $h = n$ is generated by the n-cycle $(1\ 2\ 3\ ...\ n)$.

$\mathbf{E}_n$　The *identic group* of order $h = 1$ consists of only the identical permutation $(1)\,(2)\,(3)\ ...\ (n)$.

As can be easily recognized, all permutation groups of degree n and of order less than $n!$ are proper subgroups of $\mathbf{S}_n$ (CAYLEY's theorem, see, for instance, p. 16 of [33]). For the groups listed above we have in particular:

$$\mathbf{E}_n \subset \mathbf{A}_n \subset \mathbf{S}_n . \tag{18}$$

$$\mathbf{E}_n \subset \mathbf{C}_n \subset \mathbf{D}_n \subset \mathbf{S}_n . \tag{19}$$

In general, $\mathbf{D}_n$ is not a subgroup of $\mathbf{A}_n$, except for $n = 4k + 1$. Also, $\mathbf{C}_n$ is a subgroup of $\mathbf{A}_n$ only if n is odd. For an exercise the reader may prove (18) and (19) by means of Rule 2.

For $n \leq 3$ some degeneracies in Eqs. (18) and (19) occur, namely $\mathbf{C}_3 = \mathbf{A}_3$, $\mathbf{D}_3 = \mathbf{S}_3$; $\mathbf{E}_2 = \mathbf{A}_2$, $\mathbf{C}_2 = \mathbf{D}_2 = \mathbf{S}_2$; $\mathbf{E}_1 = \mathbf{C}_1 = \mathbf{D}_1 = \mathbf{A}_1 = \mathbf{S}_1$. It is also worth noting that $\mathbf{S}_2, \mathbf{S}_3$ and $\mathbf{S}_4$ are isomorphic with $\mathbf{C}_s, \mathbf{C}_{3v}$ and $\mathbf{T}_d$, respectively, but $\mathbf{S}_5$ is not isomorphic with $\mathbf{I}_h$.

The symmetric group $\mathbf{S}_n$ is the automorphism group of the complete graph K_n as well as of its complement,

$$\mathbf{A}(K_n) = \mathbf{A}(\bar{K}_n) = \mathbf{S}_n \,. \tag{20}$$

The dihedral group $\mathbf{D}_n$ is the automorphism group of the cycle C_n as well as of its complement,

$$\mathbf{A}(C_n) = \mathbf{A}(\bar{C}_n) = \mathbf{D}_n \,. \tag{21}$$

The character tables of the symmetric groups $\mathbf{S}_n$, $n \leq 6$ are given in Appendix 5. These tables are arranged so that the classes forming the alternant group $\mathbf{A}_n$ are on the right-hand side.

9.6 Direct Product and Wreath Product

There are several group operations [34] for the generation of a group from two groups. The direct and the wreath product of two automorphism groups are of special interest. In order to explain these group operations we shall use two permutation groups $\mathbf{A} = \{A_j\}$ and $\mathbf{B} = \{B_k\}$ which operate on the set of objects $\mathcal{X} = \{x_\xi\}$ and $\mathcal{Y} = \{y_\eta\}$, respectively, such that $A_j x_\xi \in \mathcal{X}$ and $B_k y_\eta \in \mathcal{Y}$ and $\mathcal{X} \cap \mathcal{Y} = \emptyset$. The orders and the degrees of these two groups are denoted by $h(\mathbf{A})$, $h(\mathbf{B})$, $g(\mathbf{A})$ and $g(\mathbf{B})$, respectively. Note that

$$g(\mathbf{A}) = |\mathcal{X}| \quad \text{and} \quad g(\mathbf{B}) = |\mathcal{Y}| \,. \tag{22}$$

The *sum* or the *direct product* $\mathbf{A} \oplus \mathbf{B}$ is a permutation group which operates upon the union of the sets $\mathcal{X}$ and $\mathcal{Y}$. Hence its degree is $g(\mathbf{A} \oplus \mathbf{B}) = g(\mathbf{A}) + g(\mathbf{B})$. The elements of this group are ordered pairs $(A_j \oplus B_k)$ and the order of the group is $h(\mathbf{A} \oplus \mathbf{B}) = h(\mathbf{A})\,h(\mathbf{B})$. The element $z \in \mathcal{X} \cup \mathcal{Y}$ is permuted according to the rule

$$(A_j \oplus B_k)\,z = \begin{cases} A_j z & \text{if } z \in \mathcal{X}, \\ B_k z & \text{if } z \in \mathcal{Y}. \end{cases} \tag{23}$$

There are two graph operations (see Sect. 4.4) which correspond to the direct product: the union $G_a \cup G_b$ and the compound $G_a \oplus G_b$ of two graphs G_a and G_b. Thus the respective automorphism groups are obtained as

$$\mathbf{A}(G_a \cup G_b) = \mathbf{A}(G_a) \oplus \mathbf{A}(G_b); \qquad \mathbf{A}(G_a \oplus G_b) = \mathbf{A}(G_a) \oplus \mathbf{A}(G_b) \,. \tag{24}$$

Since the complete bipartite graph can be obtained as the compound of two null graphs, $K_{a,b} = \bar{K}_a \oplus \bar{K}_b$ (see Sect. 4.4), from Eqs. (20) and (24) follows

$$A(K_{a,b}) = S_a \oplus S_b \,. \tag{25}$$

The *wreath product* **A[B]** is a permutation group operating upon the cartesian product $\mathscr{X} \otimes \mathscr{Y}$ of the sets $\mathscr{X}$ and $\mathscr{Y}$. Hence the degree of the group is given by

$$g(\mathbf{A[B]}) = g(\mathbf{A})\, g(\mathbf{B}) \,. \tag{26}$$

Note that the set $\mathscr{X} \otimes \mathscr{Y}$ is composed of all ordered pairs $(x_\xi \otimes y_\eta)$, $1 \leq \xi \leq |\mathscr{X}|$, $1 \leq \eta \leq |\mathscr{Y}|$. For each $A_j \in \mathbf{A}$ and each sequence $(B_{k_1}, B_{k_2}, \ldots, B_{k_{g(\mathbf{A})}})$ of $g(\mathbf{A})$ (not necessarily different) permutations from **B**, there is in **A[B]** exactly one permutation $(A_j, B_{k_1}, B_{k_2}, \ldots, B_{k_{g(\mathbf{A})}})$ which permutes the elements $x_\xi \otimes y_\eta$ of $\mathscr{X} \otimes \mathscr{Y}$ according to the rule

$$(A_j, B_{k_1}, B_{k_2}, \ldots, B_{k_{g(\mathbf{A})}})(x_\xi \otimes y_\eta) = (A_j x_\xi) \otimes (B_\xi y_\eta) \,. \tag{27}$$

From the construction of the group elements one may conclude that

$$h(\mathbf{A[B]}) = h(\mathbf{A})\, [h(\mathbf{B})]^{g(\mathbf{A})} \,. \tag{28}$$

The relevance of the wreath product for the composition of graphs has already been mentioned in Sect. 4.4 (see Theorem 4.3). From Eq. (28) one immediately concludes that the wreath products **A[B]** and **B[A]** cannot be isomorphic.

Wreath products play an important role in the discussion of the symmetry of non-rigid molecules [54] (see Sect. 10.3) as well as in the theory of regular polymers [161].

9.7 The Representation of Automorphism Groups as Group Products

In many molecular graphs several orbits of equivalent vertices appear which may be automorphically mapped independently of each other. In such cases, the automorphism group can usually be expressed in the form of a direct product or wreath product [153]. We shall exemplify this for the graphs G_1, G_2 and G_3.

As shown in (3), the vertices 1 and 2 can only be mapped identically. They form a subset upon which the identic group E_2 operates. The two permutations of the vertices 3 and 4 form the symmetric group S_2 whereas the six permutations of the vertices 5, 6 and 7 agree with S_3. One could, therefore, expect that $A(G_1)$ is the direct product of the three groups,

$$A(G_1) = E_2 \oplus S_2 \oplus S_3 \,. \tag{29}$$

Bearing in mind the definition of the direct product (see Sect. 9.6), we see that
$$g(E_2 \oplus S_2 \oplus S_3) = 2 + 2 + 3 = 7 \quad \text{and} \quad h(E_2 \oplus S_2 \oplus S_3) = 1 \cdot (2!) \cdot (3!) = 12.$$

These results agree with the degree and order of $A(G_1)$. The correctness of Eq. (29) is definitely verified by expanding its right-hand side and comparing the result obtained with (3).

In a similar manner one obtains

$$A(G_2) = \mathbf{E}_3 \oplus \mathbf{S}_2 \oplus \mathbf{S}_2 \,. \tag{30}$$

In the case of G_3 we have once again the group $\mathbf{E}_3$ operating upon the vertices 1, 2 and 5. The vertices 3 and 6 on the one, and 4 and 7 on the other hand form certain subunits of G_3, which can be mapped onto each other. This mapping is described correctly by $\mathbf{S}_2$, but the mapping of the vertices within the subunits, $3 \leftrightarrow 3$ and $6 \leftrightarrow 6$, as well as $4 \leftrightarrow 4$ and $7 \leftrightarrow 7$, is described by $\mathbf{E}_2$. Thus the automorphic mapping of the vertices 3, 4, 6 and 7 is described by the wreath product $\mathbf{S}_2[\mathbf{E}_2]$. Therefore, one obtains

$$A(G_3) = \mathbf{E}_3 \oplus \mathbf{S}_2[\mathbf{E}_2] \,. \tag{31}$$

The degree, $g(A(G_3)) = 3 + 2 \cdot 2 = 7$, and the order, $h = 1 \cdot (2 \cdot 1^2) = 2$ of $A(G_3)$ are correctly reproduced.

Recently another approach to finding the group structure of simple graphs has been proposed [89] which makes use of the fact that the cycle structure of feasible permutations may be traced in the adjacency matrix of the graph considered. Further, attention should be drawn to the recent review of K. BALASUBRAMANIAN, Chem. Rev. *85*, 599 (1985), and the references cited therein.

Chapter 10

Some Interrelations between Symmetry and Automorphism Groups

10.1 The Idea of Rigid Molecules

Apart from the identity operation E, the elements of a space group are geometrically well-defined subspaces of the 3-dimensional space, i.e. points (i), straight lines (C_n, S_n) and planes (σ_h, σ_v, σ_d). Thus the operations associated with these elements are realized in each point of the space. On the other hand, each point of the space and, thus, each center of the molecule is transformed by such an operation, provided the point does not belong to the symmetry element considered. As a final consequence of these circumstances, the molecule treated by means of symmetry groups is considered to be rigid. This means that each atom of the molecule is associated with a triple of co-ordinates characterizing its *mean position* with absolute precision. Vibrations of the atoms about their mean positions are discussed in terms of *elongation vectors* (see paragraph 8.4.3). This concept begins to break down when an internal degree of freedom (e.g. the torsion of a methyl group) becomes fully excited. As the shall show later, automorphism groups of the molecular graph are well-suited to treat the symmetry in non-rigid molecules.

10.2 Local Symmetries

1-Cyclopentadienyl-3-cycloheptatrienylbenzene (I) has highest spatial symmetry if all the three rings are coplanar. In this case I belongs to the symmetry groups $\mathbf{C}_s = \{E, \sigma_h\}$

Q   I   P

and all π-MO's of I belong to the irreducible representation A'' of this group. Hence, the adjacency matrix (and, consequently, the HÜCKEL molecular orbital Hamiltonian) of I cannot be factorized by means of $\mathbf{C}_s$.

The molecular graph of I exhibits some *local symmetry* in the external rings, indicated by P and Q, respectively. No geometric symmetry element can be assigned to these local symmetries and, therefore, $\mathbf{C}_s$ cannot take account of them. Contrary to this, the automorphism group of I will contain permutations corresponding to P, Q and $R = PQ = QP$, i.e. $\mathbf{A}(I) = \{E, P, Q, R\}$. Thus the order of $\mathbf{A}(I)$ is higher than the order of the symmetry group, $h(A(I)) = 4 > h(C_s) = 2$. The elements of $A(I)$ have the following cycle structure:

$$\text{cs}\,(E) = [1]^{18}\,, \qquad \text{cs}\,(P) = [1]^{14}[2]^2\,, \qquad \text{cs}\,(Q) = [1]^{12}[2]^3\,,$$

$$\text{cs}\,(R) = [1]^8[2]^5\,. \tag{1}$$

Since $A(I)$ is Abelian, it possesses four irreducible representations. As a consequence of that, the adjacency matrix of I is factorized into three blocks of order 13, 3 and 2, respectively.

In the case of poly-*m*-phenylene the local symmetry of the terminal phenyl rings causes two isolated levels within the gap between the two highest valence bands of this polymer [161].

For more details concerning local symmetries see [176, 180].

10.3 Non-Rigid Molecules

Boron trifluoride (II), BF_3, is a planar molecule, that is the mean positions of the four atoms lie in one plane, σ_h. The symmetry group of BF_3 is D_{3h}, with the order $h(D_{3h}) = 12$. The automorphism group of the molecular graph of BF_3 is $A(II) = E_1 \oplus S_3$ with the order $h(A(II)) = 6$.

$$\text{II} \qquad\qquad \text{III} \qquad\qquad \text{IV}$$

The elements of $A(II)$ are:

$$E: \begin{pmatrix} 1 & 2 & 3 & 4 \\ 1 & 2 & 3 & 4 \end{pmatrix}, \qquad D: \begin{pmatrix} 1 & 2 & 3 & 4 \\ 1 & 3 & 4 & 2 \end{pmatrix}, \qquad F: \begin{pmatrix} 1 & 2 & 3 & 4 \\ 1 & 4 & 2 & 3 \end{pmatrix},$$

$$A: \begin{pmatrix} 1 & 2 & 3 & 4 \\ 1 & 2 & 4 & 3 \end{pmatrix}, \qquad B: \begin{pmatrix} 1 & 2 & 3 & 4 \\ 1 & 4 & 3 & 2 \end{pmatrix}, \qquad C: \begin{pmatrix} 1 & 2 & 3 & 4 \\ 1 & 3 & 2 & 4 \end{pmatrix}. \tag{2}$$

They are grouped in the following classes: $\{E\}$, $\{D, F\}$, $\{A, B, C\}$ [153].

Note that the symmetry group of BF_3 has higher order than its automorphism group.

The mean position of the boron and the carbon atoms of trimethylboron (III), $B(CH_3)_3$, also lie in one plane, but depending on the location of the hydrogen atoms, various symmetry groups could be appropriate. In Fig. 10.1 some of these configurations are depicted schematically. In IIIa and IIId three hydrogen atoms are located in the molecular plane, from which this plane gains the property of a symmetry element σ_h. The symmetry of IIIa is C_{3h} whereas that of IIId is only C_s. When starting with IIIa the methyl groups rotate conrotatorily in phase. i.e. $\varphi_A = \varphi_B = \varphi_C = \varphi$,

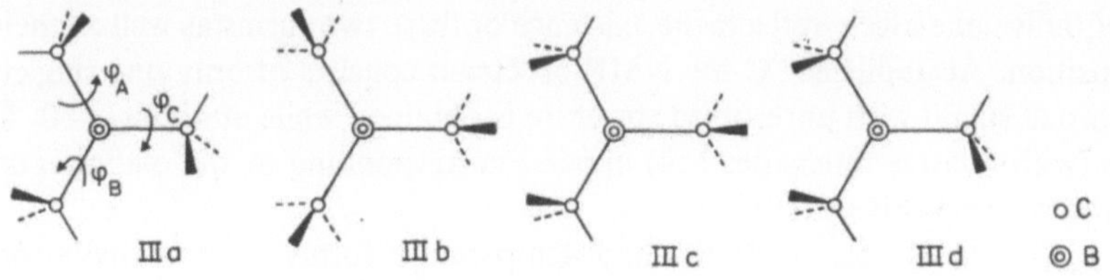

Fig. 10.1. Some particular conformations of trimethylboron (III)

then the molecule passes through various C_3 conformations and arrives with $\varphi = \pi/6$ at the configuration III b which exhibits C_{3v} symmetry. An in-phase-disrotatory motion of two methyl groups, say $\varphi_A = -\varphi_B$, $\varphi_C = 0$, transforms III b to III c, showing C_s symmetry. In contrast to III d, here only one hydrogen atom is located in the σ_h plane. In III c σ_h is perpendicular to the molecular plane while in III d these two planes coincide. Obviously, any out-of-phase motion of the methyl groups ruins all the symmetry elements and generates conformations of C_1 symmetry. Thus, the orders of the symmetry groups associable with III range between 1 and 6.

Topologically, the three methyl groups of III are equivalent. Therefore, the automorphism group of III should be similar to that of BF_3, $A(II) = E_1 \oplus S_3$. But whereas in II the S_3 operates on the three F atoms, in the case of III S_3 operates on the composed "methyl" units. Since the hydrogen atoms of a given methyl group are equivalent, the automorphism group of an isolated methyl would take the form $E_1 \oplus S_3$, where E_1 operates on the carbon and S_3 on the hydrogen atoms. In view of that and the equivalence of the methyl groups, the automorphism group of III is

$$A(III) = E_1 \oplus S_3[E_1 \oplus S_3] \, . \tag{3}$$

This expression reflects the constitution of $B(CH_3)_3$ with astonishing transparency. According to Eqs. (9.26) and (9.28), the order of $A(III)$ is $1 \cdot 6 \cdot (1 \cdot 6)^3 = 1296$ and its degree is $1 + 3 \cdot (1 + 3) = 13$.

The above analysis shows [153] that no simple symmetry group can take into account the equivalence of the hydrogens in III as indicated by its NMR spectrum [82, 91]; but $A(III)$, given by Eq. (3), does that. Concerning the treatment of the symmetry of non-rigid molecules by means of symmetry groups see [1, 71, 142, 172].

The high order of $A(III)$ is due to the automorphic mappings of the hydrogen atoms and, indeed, the subgroup $S_3[S_3]$, operating only upon the hydrogen atoms has the same order, $h(S_3[S_3]) = 1296$. The group $S_3[S_3]$ is a subgroup of S_9, which would be appropriate for nine equivalent objects if they were not grouped into three distinguishable subsets. The orders of S_9 and $S_3[S_3]$ are in the ratio 280:1.

When the automorphism groups of the complete molecular graphs are expressed by means of direct and wreath products, then usually one does not use S_n with $n > 4$. A well-known exception is the case of bullvalene, $C_{10}H_{10}$. The non-rigid high-temperature form of this molecule corresponds to the automorphism group $S_{10}[E_2]$ of the order $10! = 3\,628\,800$ and degree 20. On the other hand, the rigid low-temperature form agrees with C_{3v}, $h(C_{3v}) = 6$. The temperature-dependence of the ^{1}H-NMR spectrum of bullvalene nicely reflects the existence of these two forms as well as their mutual transition. At about 80 °C the NMR spectrum consists of only one singlet, at 15 °C a broad signal with unresolved structure is obtained while at about −80 °C two signals (with relative intensities 6:4) appear, corresponding to the olefinic and aliphatic protons, respectively [40].

Although the NMR spectra of the high-temperature forms of trimethylboron and bullvalene indicate that in both molecules under the experimental conditions all hydrogen atoms are equivalent, these equivalencies have different reasons: in tri-methylboron it is due to the full excitation of the internal torsional motion of the methyl groups, while in the case of bullvalene it is caused by very fast rearrangements of the molecule. Since a particular conformation of bullvalene can rearrange into 3 of the 10! distinct conformations, in its high-temperature form there are simultane-ously $\frac{3}{2} \cdot 10! = 5\,443\,200$ conformational equilibria present.

10.4 What Determines the Respective Orders of the Symmetry and the Automorphism Group of a Given Molecule?

For I and III the order of the automorphism group is higher than that of the respective symmetry group because the latter does not take into account the local symmetries and non-geometric equivalencies. For the planar BF_3 (II), however, where such additional features are not present, the opposite is true. Here one element of A(II) corresponds to a pair of elements of $\mathbf{D}_{3h}$. In particular, $E \in$ A(II) corresponds to E, $\sigma_h \in \mathbf{D}_{3h}$, $D, F \in$ A(II) to $2C_3$, $2S_3 \in \mathbf{D}_{3h}$ and $A, B, C \in$ A(II) to $3C_2'$, $3\sigma_v \in \mathbf{D}_{3h}$. This is a result of the inability of the elements of A(II) to recognize the direction of a vector parallel to the z-axis, i.e. to distinguish between $+z$ and $-z$.

Consider an unbranched molecule like acetylene, $HC \equiv CH$, (IV). In addition to the identical mapping only one more automorphism exists which maps the terminal vertices, their neighbours, etc., pairwise onto each other. Hence, $h(A(IV)) = 2$. On the other hand, the symmetry group of IV is $\mathbf{D}_{\infty h}$, having one ∞-fold axis of symmetry. Consequently, $h(\mathbf{D}_{\infty h}) = \infty$. In order to understand this we must remember that the full orthogonal group $\mathbf{O}(3)$, which possesses the complete symmetry of the sphere, consists of an infinite number of ∞-fold axes of symmetry, in accordance with the high symmetry of the 3-dimensional Euclidean space. When within this space a given topology is realized, the high symmetry of $\mathbf{O}(3)$ is broken. In the case that an unbranched topology is realized by a collinear geometry, only one of the infinite number of ∞-fold axes of symmetry is conserved. Thus the high symmetry of a collinear molecule reflects rather the symmetry of the Euclidean space than the symmetry of its molecular topology, as expressed by the automorphism group of the molecular graph.

Part D

Special Topics

Chapter 11

Topological Indices

Whereas the topology of a molecule, represented by the molecular graph is an essentially non-numerical mathematical object, various measurable properties of molecules are usually expressed by means of numbers. In order to link molecular topology to any real molecular property one must first convert the information contained in the molecular graph into a numerical characteristic. Every number which is uniquely determined by a graph is called a graph invariant. Those invariants of molecular graphs which are used for structure-property or structure-activity correlations are usually called topological indices (of the corresponding molecule).

All topological indices $I(G)$ must meet the following requirement:

$$\text{if } G_1 \simeq G_2, \quad \text{then} \quad I(G_1) = I(G_2). \tag{1}$$

The reverse of this statement is by no means true. If $I(G_1) = I(G_2)$, then the graphs G_1 and G_2 need not be isomorphic.

Much effort has been made to design topological indices, or pairs, triplets etc. of topological indices which would characterize a molecular graph up to isomorphism[1]. It is commonly believed in graph theory that a non-trivial characterization of this type is not possible.

A plethora of topological indices has been proposed in the chemical and pharmaceutical literature [7, 9, 47], with no sign that their proliferation will stop in the near future. The great and increasing number of topological indices indicates that perhaps a clear and unambiguous criterion for their selection and verification is still missing. Moreover, many of the currently used topological indices are mutually correlated. Motoc and Balaban have demonstrated this for 17 indices on the example of the octane isomers [147]. On the other hand, the success in predicting certain physical, chemical and pharmacological properties of organic molecules by means of topological indices cannot be denied.

It is not the aim of this section to either review these numerous topological indices or to elaborate on their applications in practice. We shall restrict our attention to only two of them and examine some of their mathematical properties. It is somewhat surprising that, although the work on topological indices comprises a good part of research in contemporary mathematical chemistry, rather few general results (theorems) have been obtained so far.

---

[1] This so-called "isomorphism disease" seems to be widespread among beginners and amateurs in graph theory.

The first topological index was introduced in 1947 by HARRY WIENER [179] and used for correlation with boiling points of alkanes. WIENER's index is related to the distances in molecular graphs and will be considered in Sect. 11.1. Another topological index whose mathematical properties are relatively well investigated is that of HOSOYA, which will be presented in Sect. 11.2.

11.1 Indices Based on the Distance Matrix

11.1.1 The Wiener Number and Related Quantities

The *distance* between two vertices in a (connected) graph as well as the *distance matrix* have been defined in Sect. 4.1. If $v_1, v_2, \dots, v_n$ are the vertices of a connected graph G and $d(v_r, v_s)$ denotes the distance between v_r and v_s, then the elements of the distance matrix are given by

$$d_{rs} = d(v_r, v_s) . \tag{2}$$

The r-th row of the distance matrix is called the *distance vector* $\boldsymbol{D}_{v_r}$ of the vertex v_r. Hence

$$\boldsymbol{D}_{v_r} = (d_{r1}, d_{r2}, \dots, d_{rn}) . \tag{3}$$

The sum of all the entries of $\boldsymbol{D}_v$ is called the *distance number* of the vertex v and will be denoted by $d(v) = d(v, G)$.

The WIENER *number* $W = W(G)$ of the graph G is equal to the sum of the distances between all pairs of vertices of G. It is immediately clear that

$$W = \sum_{r<s} d_{rs} = \frac{1}{2} \sum_{v \in \mathscr{V}} d(v) . \tag{4}$$

A *distance tree* $T_v(G)$ of the vertex v of the graph G is a spanning tree of G with the property that $T_v(G)$ and G have identical distance vectors corresponding to the vertex v.

The existence of at least one distance tree for each vertex of a connected graph G is evident from the following construction [158].

Let $\mathscr{V}$ be the vertex set of G and let $v \in \mathscr{V}$. Partition the vertices of G into classes, such that the k-th class contains all vertices $u \in \mathscr{V}$ for which $d(u, v) = k$.

These vertices are said to form the k-th sphere of the vertex v.

1. Start with a collection of n isolated vertices whose labels coincide with those from $\mathscr{V}$.
2. Link the vertex v with all those vertices which are adjacent to v in G (the vertices of the first sphere).
3. For $k = 2, 3, \dots, n - 1$ link the vertices of the $(k - 1)$-th sphere to those vertices of the k-th sphere which are neighbours in G, but without forming any cycle.

The construction of distance trees is illustrated in Fig. 11.1.

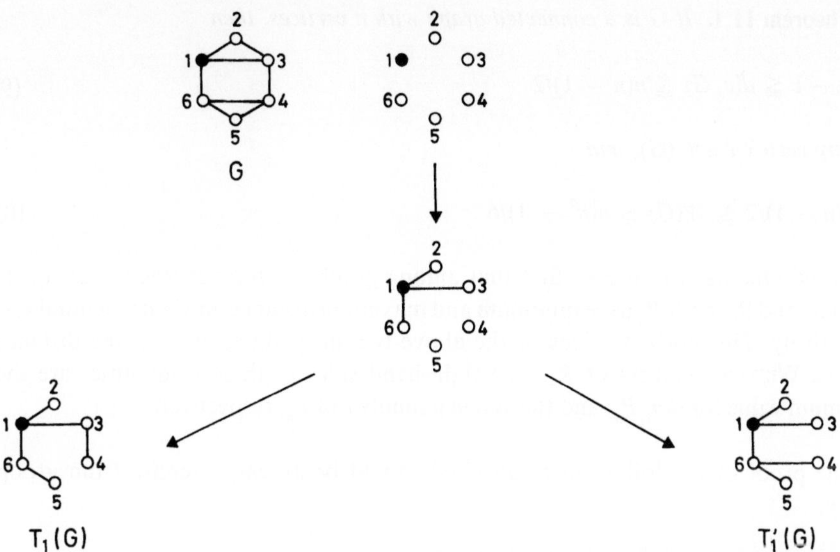

Fig. 11.1. The construction of the distance tree for the vertex 1 of the graph G. Note that in the third step either the edge e_{34} or the edge e_{46}, but not both e_{34} and e_{46} are introduced, resulting thus in two non-isomorphic trees $T_1(G)$ and $T_1'(G)$

In the general case several non-isomorphic distance trees may exist for the same vertex. The presence of even-membered cycles in G is a necessary, but not a sufficient condition for the existence of several non-isomorphic distance trees.

In order to illustrate the above definitions, consider the cycle C_n. Since all the vertices of C_n are equivalent, all its distance trees must be isomorphic. It is easy to see that $T_v(C_n) \simeq P_n$ for all $v \in \mathcal{V}(C_n)$.

By an appropriate labeling of the vertices we can obtain the distance vector of a vertex v of C_n in the form

$$D_v(C_n) = (0, 1, 2, \dots , (n-4)/2, (n-2)/2, (n-4)/2, \dots , 2, 1) \tag{5}$$

if n is even, and

$$D_v(C_n) = (0, 1, 2, \dots , (n-3)/2, (n-1)/2, (n-1)/2, (n-3)/2, \dots , 2, 1) \tag{6}$$

if n is odd. It follows now immediately that for all $v \in \mathcal{V}(C_n)$,

$$d(v, C_n) = \begin{cases} n^2/4 & \text{if } n \text{ is even} \\ (n^2 - 1)/4 & \text{if } n \text{ is odd}, \end{cases} \tag{7}$$

and consequently

$$W(C_n) = \begin{cases} n^3/8 & \text{if } n \text{ is even} \\ n(n^2 - 1)/8 & \text{if } n \text{ is odd}. \end{cases} \tag{8}$$

Theorem 11.1. *If G is a connected graph with n vertices, then*

$$n - 1 \leqq d(v, G) \leqq n(n - 1)/2 \tag{9}$$

for any vertex $v \in \mathscr{V}(G)$, and

$$n(n - 1)/2 \leqq W(G) \leqq n(n^2 - 1)/6 . \tag{10}$$

Proof. One has to observe first that among graphs with n vertices the complete graph K_n and the path P_n have minimum and maximum distance and WIENER numbers, respectively. The left-hand sides of the above two inequalities are just the distance and the WIENER numbers of K_n. The right-hand sides of these inequalities are the maximum value for $d(v, P_n)$ and the WIENER number of P_n, respectively. $\square$

The proof of the following result [158] should be an easy exercise from group theory.

Theorem 11.2. *If two vertices of a graph are automorphically mapped onto each other, then their distance numbers are equal.*

Let u and v be two adjacent vertices of the graph G. The vertex set of G can be partitioned into three classes:

$$\mathscr{V}_u = \{p \in \mathscr{V}(G) | d(u, p) < d(v, p)\}$$
$$\mathscr{V}_{uv} = \{p \in \mathscr{V}(G) | d(u, p) = d(v, p)\} \tag{11}$$
$$\mathscr{V}_v = \{p \in \mathscr{V}(G) | d(u, p) > d(v, p)\} .$$

Denote the number of elements of these classes by $|\mathscr{V}_u|$, $|\mathscr{V}_{uv}|$ and $|\mathscr{V}_v|$, respectively. Without proof we state the following result [158].

Theorem 11.3.

$$d(u, G) = d(v, G) + |\mathscr{V}_v| - |\mathscr{V}_u| . \tag{12}$$

It is interesting to note that the set $\mathscr{V}_{uv}$ is non-empty if and only if the vertices u and v belong to an odd-membered cycle.

Theorem 11.3 provides an efficient method for the calculation of the distance number and thus also the WIENER index of a graph. The procedure requires the knowledge of a single distance number (which can be conveniently achieved by constructing a single distance tree). The distance numbers of all other vertices of G can be then determined step-by-step using Eq. (12).

In the case of trees, there is an even more rapid way for the calculation of the WIENER number [179].

Theorem 11.4. *Let T be a tree and use the same notation as in Theorem 11.3. Then*

$$W(T) = \sum_{u,v} |\mathscr{V}_u| \cdot |\mathscr{V}_v| \tag{13}$$

with the summation going over all pairs of adjacent vertices of T.

Proof. By Theorem 6.3, any pair of vertices in a tree is connected by a unique path. The number of edges of this path is just the distance between the respective two vertices.

Now, in the case of trees the Wiener index can be obtained either be summing the distances between all pairs of vertices or summing the number of paths which contain a given edge, over all edges.

Let e be an edge of T, connecting the vertices u and v. Then for all $p \in \mathscr{V}_u$ and $q \in \mathscr{V}_v$ the (unique) path connecting p and q contains e. The total number of paths containing e is thus $|\mathscr{V}_u| \cdot |\mathscr{V}_v|$ and Theorem 11.4 follows immediately. $\square$

11.1.2 Applications of the Wiener Number

As already mentioned, the Wiener number is the first in a long series of topological indices. In his pioneering paper [179] Wiener introduced not only the "path number" (which is nowadays known under his name), but also the "polarity number" — the number of selections of P_4 subgraphs in the molecular graph. By using a linear combination of the "path number" and the "polarity number", Wiener was able to give a reasonably reliable prediction of boiling points of alkanes. In a series of subsequent papers, Wiener and independently Platt extended the application of these indices to a number of other physico-chemical properties of alkanes (heats of formation, heats of vaporization, molar volumes and molar refractions). Since the Wiener number measures the compactness of a molecule (c.f. Theorem 11.1), it is clear that it can be correlated with those physico-chemical properties which depend on the volume/surface ratio of the molecules. Gas-chromatographic retention data for series of structurally related molecules are typical examples where the Wiener number "works". Less common applications of the Wiener number have also been reported, e.g. in modeling crystal growth and crystal vacancies and certain properties of conjugated polymers. For further data on the applications of the Wiener number and an extensive bibliography, the interested reader should consult the review [7].

11.2 Hosoya's Topological Index

Hosoya [129] in 1971 seems to be the first to conceive the chemical significance of certain combinatorial properties of the molecular graphs of saturated hydrocarbons, namely those related to the selection of mutually nonadjacent (or as graph theoreticians use to say: independent) edges. Whereas the perfect matchings of molecular graphs (called Kfkulé structures), have attracted the attention of chemists over a whole century, the non-perfect matchings have been long disregarded.

Hosoya introduced the quantities $m(G, k)$ as the number of selections of k mutually nonadjacent CC bonds in the hydrocarbon whose molecular graph is G and called their sum the topological index. It was claimed that this index was a candidate for classifying saturated hydrocarbons with respect to their topological nature. Although it is not capable of reflecting the entire structure of isomers, it is roughly dependent on their size, mode of branching and ring closure [129].

11.2.1 Definition and Chemical Applications of Hosoya's Index

The number of k-matchings in a graph, denoted by $m(G, k)$, was defined in paragraph 4.2.2 where its basic mathematical properties were also described. Hence, the topological index defined in [129] is given as

$$Z = Z(G) = \sum_{k=0}^{[n/2]} m(G, k) \tag{14}$$

and is thus equal to the total number of matchings of the graph G. The graph G was originally assumed to represent the carbon-atom skeleton of a saturated hydrocarbon. There is, however, no obstacle to considering $Z(G)$ for an arbitrary graph $G \in \mathcal{G}$.

$Z(G)$ is usually called Hosoya's topological index. The descriptor "topological" will be dropped in the following text.

It has been shown that a nearly linear correlation exists between the logarithm of Z and the boiling point of the corresponding saturated hydrocarbon [129, 131]. The quality of this correlation is illustrated on Fig. 11.2.

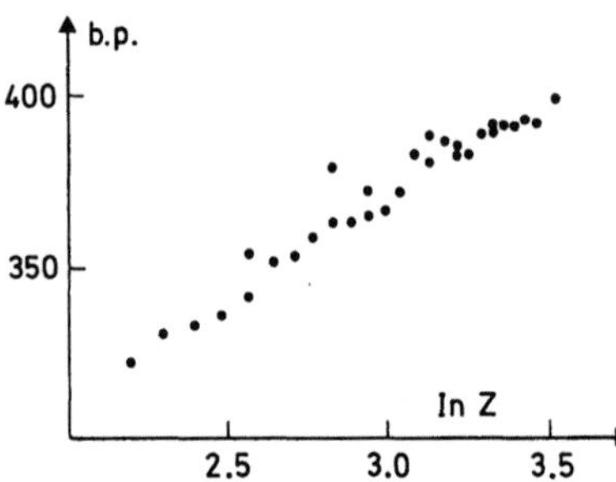

Fig. 11.2. Boiling points (in K at normal pressure) of alkanes with 6, 7 and 8 carbon atoms versus the logarithm of Hosoya's index

A better reproduction of boiling points was gained by the formula $(a \ln Z + b) n^{-1/2} + c$, where a, b and c are empirical parameters [131].

In a later work it was found that the absolute entropy of acyclic saturated hydrocarbons correlates well with the logarithm of Z, except for sterically overcrowded and highly symmetric molecules. The physical meaning of this relation was clarified by analyzing the rotational partition function [152].

Figure 11.3 should provide an illustration of the connection between entropy and Hosoya's index.

Attempts to use the index Z for classifying structural formulas in chemical documentation were also reported [130].

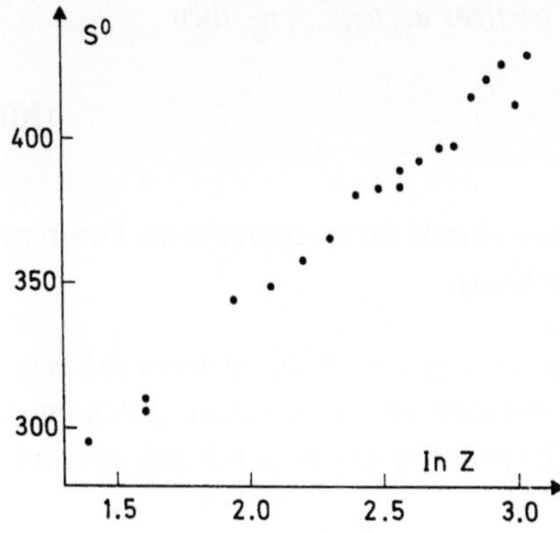

Fig. 11.3. Entropy (in $JK^{-1}mol^{-1}$ at 298 K) of alkanes with 4 to 7 carbon atoms versus the logarithm of Hosoya's index

11.2.2 Mathematical Properties of Hosoya's Index

In order to make the manipulations with $Z(G)$ easier, it would be convenient to introduce a polynomial

$$\alpha^\dagger(G, x) = \sum_{k=0}^{[n/2]} m(G, k)\, x^k \tag{15}$$

which is in fact the generating function for the number of k-matchings of the graph G. Of course, $Z(G) = \alpha^\dagger(G, 1)$.

It is clear that the polynomial $\alpha^\dagger(G)$ is closely related to the matching polynomial $\alpha(G)$, see Eq. (4.11). The explicit relation between them is

$$\alpha(G, x) = x^n \alpha^\dagger(G, -x^{-2}) \tag{16}$$

or

$$\alpha^\dagger(G, x) = (i/\sqrt{x})^n\, \alpha(G, i/\sqrt{x}) \tag{17}$$

where $i = \sqrt{-1}$.

Consequently, there is no need to elaborate the properties of $\alpha^\dagger(G)$. All we have to do is to directly reformulate the results known for the matching polynomial.

As a matter of fact, the recurrence relations for the matching polynomial are fully analogous to the recurrence relations for the characteristic polynomial of trees, except that they hold for all graphs. We shall summarize them in a theorem.

Theorem 11.5. *(a) If e is any edge of a graph $G \in \mathcal{G}$, connecting the vertices u and v, then*

$$\alpha(G) = \alpha(G - e) - \alpha(G - u - v)\,. \tag{18}$$

(b) If v is a pendent vertex of G, being adjacent to the vertex u, then

$$\alpha(G) = x\alpha(G - v) - \alpha(G - u - v)\,. \tag{19}$$

(c) If v is a vertex of G, being adjacent to the vertices $u_1, u_2, \ldots, u_g$, then

$$\alpha(G) = x\alpha(G - v) - \sum_{i=1}^{g} \alpha(G - u_i - v) . \tag{20}$$

The parts (a), (b) and (c) of the above theorem should be compared with Theorem 6.5, Corollary 6.5.1 and Corollary 6.5.2, respectively.

Proof. All the statements given in Theorem 11.5 can be deduced from the relations (4.9) and (4.10), bearing in mind the definition of the matching polynomial (4.11). The details are similar to those used in the proof of Theorem 6.5 and its corollaries and will be omitted here. $\square$

Recursion relations for HOSOYA's index are now a straightforward consequence of Theorem 11.5 and the fact that

$$Z(G) = (--i)^n \alpha(G, i) . \tag{21}$$

This latter formula follows, of course, from Eq. (17).

Theorem 11.6. *(a) If e is any edge of a graph $G \in \mathcal{G}$, connecting the vertices u and v, then*

$$Z(G) = Z(G - e) + Z(G - u - v) . \tag{22}$$

(b) If v is a pendent vertex of G, being adjacent to the vertex u, then

$$Z(G) = Z(G - v) + Z(G - u - v) . \tag{23}$$

(c) If v is a vertex of G, being adjacent to the vertices $u_1, u_2, \ldots, u_g$, then

$$Z(G) = Z(G - v) + \sum_{i=1}^{g} Z(G - u_i - v) . \tag{24}$$

For the actual calculation of $Z(G)$ the following simple result is rather useful. If G is composed of components $G_1, G_2, \ldots, G_k$, then

$$Z(G) = Z(G_1) Z(G_2) \ldots Z(G_k) . \tag{25}$$

As an example we shall calculate HOSOYA's index for the tricyclic graph $G(n, i, j)$

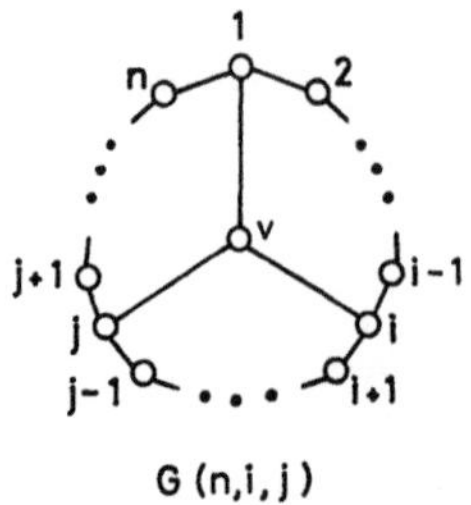

G (n, i, j)

Applying (24) to the central vertex v of $G(n, i, j)$ we get

$$Z(G(n, i, j)) = Z(C_n) + 3Z(P_{n-1}) \qquad (26)$$

since the graph obtained by the deletion of v and any of its neighbours is isomorphic to the path with $n - 1$ vertices. Explicit expressions for $Z(C_n)$ and $Z(P_{n-1})$ will be deduced in the subsequent paragraph.

It is worth noting that $Z(G(n, i, j))$ does not depend on the parameters i and j. As a consequence, systems like $G(12, 5, 9)$, $G(12, 5, 10)$, $G(12, 3, 7)$ etc. are indistinguishable as long as the Z index is considered.

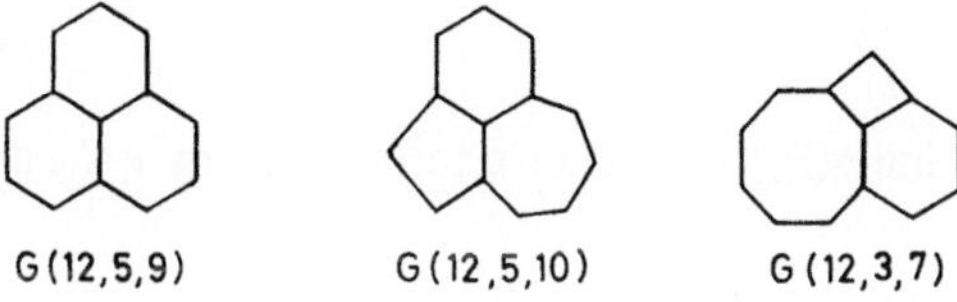

G (12,5,9)   G (12,5,10)   G (12,3,7)

11.2.3 Example: Hosoya's Index of the Path and the Cycle

In order to make the reader a bit more familiar with Hosoya's index we shall work out two further examples, exhibiting certain interesting combinatorial features of $Z(G)$.

Corollary 11.6.1. *For the path P_n with n vertices, $Z(P_n)$ is equal to the n-th* Fibonacci *number. Therefore,*

$$Z(P_n) = \left[[(1 + \sqrt{5})/2]^{n+1} - [(1 - \sqrt{5})/2]^{n+1} \right] \Big/ \sqrt{5} . \qquad (27)$$

Proof. Recall that the Fibonacci numbers F_n, $n = 2, 3, 4, \ldots$ are generated recursively from the relation $F_n = F_{n-1} + F_{n-2}$, using the initial conditions $F_0 = F_1 = 1$. Now, because of (6.22),

$$Z(P_n) = Z(P_{n-1}) + Z(P_{n-2}) \qquad (28)$$

which is just the Fibonacci recursion. By direct calculation we can check that $Z(P_n)$ coincides with F_n for, say, $n = 1, 2$ and 3. Then $Z(P_n)$ will coincide with F_n for all n.

The expression (27) for $Z(P_n)$ is the well known Binét formula for the Fibonacci numbers. $\square$

Corollary 11.6.2. *For the cycle C_n with n vertices, $Z(C_n)$ is equal to the n-th* Lucas *number. Therefore,*

$$Z(C_n) = [(1 + \sqrt{5})/2]^n + [(1 - \sqrt{5})/2]^n . \qquad (29)$$

Proof. We shall first show that $Z(C_n)$ obeys a FIBONACCI-type recursion relation, namely

$$Z(C_n) = Z(C_{n-1}) + Z(C_{n-2}) \,. \tag{30}$$

For this, it is sufficient to verify that

$$\alpha(C_n) = x\alpha(C_{n-1}) - \alpha(C_{n-2}) \,. \tag{31}$$

Apply Theorem 11.5a to any edge of the cycle. This gives

$$\alpha(C_n) = \alpha(P_n) - \alpha(P_{n-2}) \,. \tag{32}$$

Using (6.32) and having in mind that $\alpha(P_n) = \varphi(P_n)$ we can transform $\alpha(C_n)$ as:

$$\begin{aligned}
\alpha(C_n) &= [x\alpha(P_{n-1}) - \alpha(P_{n-2})] - [x\alpha(P_{n-3}) - \alpha(P_{n-4})] \\
&= x[\alpha(P_{n-1}) - \alpha(P_{n-3})] - [\alpha(P_{n-2}) - \alpha(P_{n-4})] \,.
\end{aligned} \tag{33}$$

By (32), $\alpha(P_{n-1}) - \alpha(P_{n-3}) = \alpha(C_{n-1})$ and $\alpha(P_{n-2}) - \alpha(P_{n-4}) = \alpha(C_{n-2})$ and one arrives at (31).

Of course, the relation (30) could also be obtained by a direct application of Theorem 11.6.

The rest of the proof is now completely analogous to that of Corollary 11.6.1. Recall that the LUCAS numbers L_n, $n = 1, 2, \dots$ are defined recursively via $L_n = L_{n-1} + L_{n-2}$ with the initial conditions $L_0 = 2$, $L_1 = 1$. $\square$

11.2.4 Some Inequalities for Hosoya's Index

The quasiordering of graphs introduced in paragraph 4.1.4 has a straightforward application to HOSOYA's index. Namely, if $G_1 \succ G_2$, then also $Z(G_1) \geqq Z(G_2)$ with equality only if the graphs G_1 and G_2 are matching equivalent.

From Theorem 6.6 we can now immediately conclude that among trees with n vertices the star has the minimum and the path the maximum value for Z. The following statement is somewhat stronger. Let T_a, T_b, T_c and T_d be trees with n vertices:

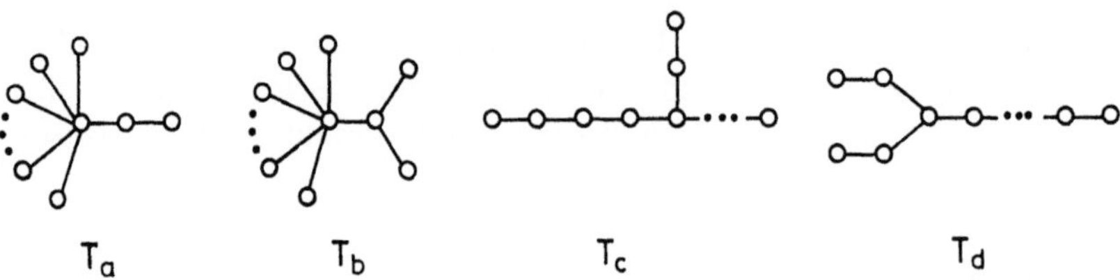

T_a T_b T_c T_d

Theorem 11.7. If T is any tree from the set $\mathcal{T}_n \setminus \{K_{1,n-1}, T_a, T_b, T_c, T_d, P_n\}$, then

$$Z(K_{1,n-1}) < Z(T_a) < Z(T_b) < Z(T) < Z(T_c) < Z(T_d) < Z(P_n) \,. \tag{34}$$

This statement applies for $n \geq 8$. For smaller values of n some of the above inequalities need to be slightly modified.

Numerous relations for Hosoya's index follow from the results given in [100, 104, 113] (see paragraph 6.1.4). We present here only one of them, which has a direct experimental consequence.

Let the tree $P_n(j)$ be obtained by joining a new (pendent) vertex to the j-th vertex of the path P_n.

Note that $P_n(1) \simeq P_n(n) \simeq P_{n+1}$.

Theorem 11.8. *Let $n = 4k - 1$ or $4k$ or $4k + 1$ or $4k + 2$. Then*

$$Z(P_{n+1}) \geq Z(P_n(3)) \geq Z(P_n(5)) \geq \ldots \geq Z(P_n(2k - 1)) \geq Z(P_n(2k + 1))$$

$$\geq Z(P_n(2k)) \geq Z(P_n(2k - 2)) \geq \ldots \geq Z(P_n(4)) \geq Z(P_n(2)) \, . \tag{35}$$

Proof. Apply Theorem 11.6b to the third pendent vertex of $P_n(j)$. This gives

$$Z(P_n(j)) = Z(P_n) + Z(P_{j-1} \cup P_{n-j}) \, . \tag{36}$$

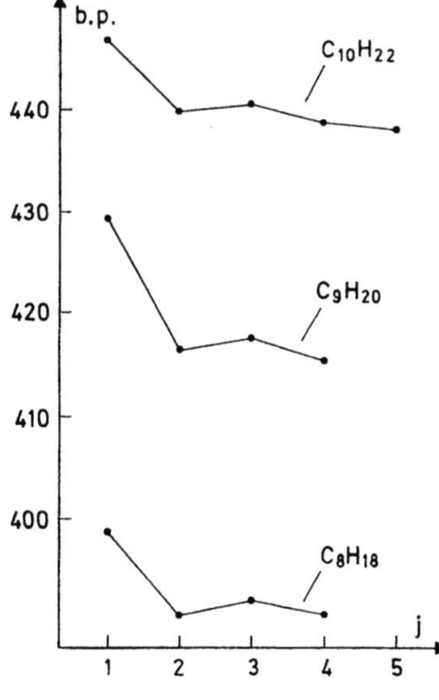

Fig. 11.4. Boiling points (in K) of monomethylalkanes as a function of the position (j) of the methyl group. Note that $j = 1$ corresponds to the normal alkanes

The first summand on the right-hand side is independent of j. Therefore, $Z(P_n(j)) \geqq Z(P_n(j^*))$ whenever $Z(P_{j-1} \cup P_{n-j}) \geqq Z(P_{j^*-1} \cup P_{n-j^*})$. This latter will occur if $P_{j-1} \cup P_{n-j} > P_{j^*-1} \cup P_{n-j^*}$ and Theorem 11.8 is obtained from Lemma 6.7. $\square$

The tree $P_n(j)$ is, however, the molecular graph of a monomethylalkane (e.g. $P_7(3)$ represents the 3-methylheptane). Therefore, Theorem 11.8 implies certain regularities in the boiling points and entropies of monomethylalkanes. In particular, the boiling points should alternately decrease and increase as the methyl group is "moved" along the carbon-atom skeleton. How good the experimental findings follow the predictions based on Theorem 11.8 can be seen from Fig. 11.4.

Chapter 12

Thermodynamic Stability of Conjugated Molecules

12.1 Total π-Electron Energy and Thermodynamic Stability of Conjugated Molecules

We already know that the molecular orbital energy levels of the π-electrons in conjugated molecules are (within the HMO approximation) related to the eigenvalues of the molecular graph via

$$E_j = \alpha + \lambda_j \beta . \tag{1}$$

If the number of π-electrons in the j-th molecular orbital is η_j (where, of course, η_j is either 2 or 1 or 0), then the total energy of all π-electrons in the molecule considered is given by

$$E_\pi = \sum_{j=1}^{n} \eta_j E_j . \tag{2}$$

Combining (1) and (2) we immediately arrive at

$$E_\pi = n_e \alpha + \beta \sum_{j=1}^{n} \eta_j \lambda_j \tag{3}$$

where n_e denotes the number of π-electrons. The term $n_e \alpha$ corresponds to the energy of n_e isolated p-electrons. The second term on the right-hand side of (3) is the energy gained from the interaction of these p-electrons in the molecule and is sometimes called the π-binding energy. Since β is a constant, the only non-trivial part of (3) is

$$E = \sum_{j=1}^{n} \eta_j \lambda_j . \tag{4}$$

The quantity E, defined via (4) is also called the HMO total π-electron energy and will be the object considered in the present chapter. It is formally obtained from (3) by using the so-called β-units: $\alpha = 0$ and $\beta = 1$. One should, however, always have in mind that β is a negative constant[1] (whose actual numerical value is irrelevant for the problems with which we shall be concerned), so that the total (electron) energy of a conjugated molecule decreases with increasing E. Hence, the larger is the value of E the greater is the (predicted) thermodynamic stability of the corresponding compound.

---

[1] The value $\beta = -137.0 \, kJ/mol = -1.4199 \, eV$ was recommended for thermodynamic applications of the HMO model [173].

A serious objection to Eq. (3) is that it does not take into account the interactions between the π-electrons, as required by more sophisticated molecular orbital approaches. Arguments have been offered supporting the opinion that (at least a part of) the electron interaction is contained in Eq. (3) [133, 143, 170, 174]. Their consideration would go much beyond the scope of the present book. Here we rather emphasize the empirically tested fact [173] that the total π-electron energy is in good linear correlation with experimental heats of formation of conjugated hydrocarbons, especially when there is no steric strain. Heats of atomization computed by the HMO method are found to be accurate to 0.1 %, implying accuracy of $\pm 0.005\beta$ in resonance energies per π-electron.

Figure 12.1 should give a general impression about the thermodynamic relevance of the HMO total π-electron energy. More details along these lines are given in Sect. 12.9.

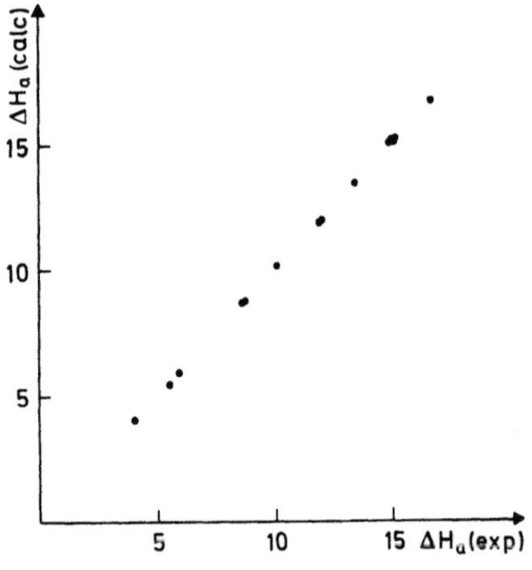

Fig. 12.1. Experimental heats of atomization (in *MJ/mol*) vs. heats of atomization calculated by the HMO method [173]

It has been recently demonstrated [133] that the HMO total π-electron energy is in a perfect linear correlation with the kinetic energy of π-electrons as calculated by rather accurate (STO-3G) ab initio methods.

One of the main uses of total π-electron energy is the calculation of resonance energy [90]. HESS and SCHAAD have demonstrated that the resonance energies calculated from HMO total π-electron energies are of equal quality and reliability as those obtained by far more advanced SCF MO methods (see [127, 128] and the numerous subsequent publications). For more detail on this matter see Sect. 12.9.

We may conclude that despite its admitted limitations, the HÜCKEL molecular orbital total π-electron energy is usually a good estimate for a given conjugated compound, when compared to more sophisticated methods. Moreover, trends within a class of topologically related compounds are generally well-accounted for.

12.2 Total π-Electron Energy and Molecular Topology

It is clear from (4) that the total π-electron energy is uniquely determined by the topology of the corresponding molecule (via the eigenvalues of its molecular graph), provided the occupation numbers η_j are fixed. This is indeed the case if we consider the conjugated molecules in their ground state. Then we immediately have

$$E = 2 \sum_{j=1}^{n_e/2} \lambda_j \tag{5}$$

if n_e is even, and

$$E = \lambda_{(n_e+1)/2} + 2 \sum_{j=1}^{(n_e-1)/2} \lambda_j \tag{6}$$

if n_e is odd, where n_e is the number of π-electrons.

In former times numerous attempts have been made to design calculation schemes for estimating E from some (easily obtainable) topological invariants of the molecule. With the advent of the era of computers, exact E values become so easily available, that any further effort along these lines turned out to be meaningless.

Another, somewhat related question continues, however, to attract the attention of contemporary researchers. This is the problem of how *E depends on the topology of the molecule*. In other words we may ask *which topological invariants of the conjugated molecule or which invariants of the molecular graph determine the total π-electron energy*. The answer to this question would give us insight into the perplexed relations which exist between the chemical structure of a conjugated molecule and its thermodynamic stability.

Fortunately the HMO model is simple enough to allow the treatment of the above stated problem. On the other hand, the dependence of the HMO total π-electron energy on molecular topology is complicated enough to enable the formulation of a variety of non-trivial mathematical statements. These will be exposed in what follows.

12.3 The Energy of a Graph

In order to avoid any misunderstanding we stress immediately that a graph (being a mathematical object) cannot possess any kind of real (physical) energy. Nevertheless, as explained in detail in the previous two sections, the binding energy of the π-electrons can be calculated from the molecular graph. In all real chemical applications, the graph from which this energy is calculated must correspond to a chemically meaningful π-electron network and is therefore subject to drastic limitations (see Chap. 3). On the other hand, the vast majority of the results which will be exposed in the present chapter hold irrespective of all such limitations. This gives us a motivation to introduce the concept of the *energy of a graph* as a proper (but formal mathematical) generalization of what is actually calculated for molecular graphs [25]. The advantage of this generalized graph energy is that most of our mathematical results will now apply to all graphs.

In the great majority of (though certainly not in all) conjugated π-electron systems of chemical interest all bonding molecular orbitals are doubly occupied and all antibonding molecular orbitals are unoccupied. (For the calculation of E the number of π-electrons in non-bonding molecular orbitals is irrelevant.) This means that $\eta_j = 2$ if $\lambda_j > 0$ and $\eta_j = 0$ if $\lambda_j < 0$, and Eqs. (5) and (6) can be written in the form

$$E = 2 \sum_{+} \lambda_j \tag{7}$$

where $\sum_{+}$ means summation over all positive graph eigenvalues.

For instance, Eq. (7) holds for all alternant hydrocarbons because of the pairing theorem (Theorem 6.9).

Even if some of the bonding MO's are empty or some antibonding MO's are filled, Eq. (7) still holds, but only as a good approximation. This is because the graph eigenvalues corresponding to the highest bonding and the lowest antibonding MO are usually negligibly small compared to E.

In all the considerations which follow, the validity of Eq. (7) will be assumed.

We prove now an elementary yet rather important consequence of (7).

Theorem 12.1. *For conjugated hydrocarbons whose total π-electron energy satisfies (7),*

$$E = \sum_{j=1}^{n} |\lambda_j| \, . \tag{8}$$

Proof. The sum of the eigenvalues of a simple graph is zero (see Eq. (4.24)). Hence

$$\sum_{+} \lambda_j + \sum_{-} \lambda_j = 0 \tag{9}$$

where $\sum_{-}$ symbolizes summation over all negative eigenvalues. Substituting the above identity into (7) we get

$$E = \sum_{+} \lambda_j - \sum_{-} \lambda_j \tag{10}$$

from which (8) is obvious. $\square$

Corollary 12.1.1. *For heteroconjugated molecules whose total π-electron energy satisfies (7),*

$$E = \sum_{v \in \mathscr{V}} h_v + \sum_{j=1}^{n} |\lambda_j| \tag{11}$$

where h_v is the weight associated with the vertex v (see Sect. 6.5).

Theorem 12.1 is important because it relates the total π-electron energy to an interesting though not simple function of the graph eigenvalues. It is, for example, worth noting that the right-hand side of (8) is a symmetric function in the sense that it remains unchanged if any two eigenvalues exchange places.

The function on the right-hand side of (8) has a certain mathematical beauty. Curiously enough, it has not previously been considered by mathematicians. This quantity is, of course, well defined for all graphs. Having in mind the previously given explanation, we shall call it *the energy of the graph* [25].

Hence let G be an arbitrary graph and $\lambda_1, \lambda_2, \ldots, \lambda_n$ its eigenvalues. Then the energy of G is defined to be equal to $|\lambda_1| + |\lambda_2| + \ldots + |\lambda_n|$ and will be denoted by $E(G)$ or (where there is no danger of misunderstanding) simply by E.

Most of the results which will be presented in this chapter are, in fact, statements about the energy of graphs.

12.4 The Coulson Integral Formula

A classical result in the topological theory of total π-electron energy is that obtained by COULSON as early as 1940 [78]. It provides a direct analytical connection between the characteristic polynomial and the energy of a graph.

Theorem 12.2. *Let G be a graph with n vertices and $\varphi'(G, x)$ the first derivative of its characteristic polynomial. Let further $i = \sqrt{-1}$. Then*

$$E(G) = \frac{1}{\pi} \int_{-\infty}^{\infty} \left[n - ix\varphi'(G, ix)/\varphi(G, ix) \right] dx$$

$$= \frac{1}{\pi} \int_{-\infty}^{+\infty} \left[n - x(d/dx) \ln \varphi(G, ix) \right] dx \ . \tag{12}$$

Here and later, by the integral $\int_{-\infty}^{+\infty} F(x)\, dx$ we always mean its proper value $\lim_{z \to \infty} \int_{-z}^{z} F(x)\, dx$.

Proof. The above result is easily obtained by contour integration [78]. We offer here another, somewhat more elementary and straightforward derivation of (12), based on the observation that

$$\varphi'(G, x)/\varphi(G, x) = \sum_{j=1}^{n} \frac{1}{x - \lambda_j} \ , \tag{13}$$

Let us consider the simple integrals

$$I_1 = \frac{1}{\pi} \int_{-\infty}^{+\infty} \frac{\lambda^2}{\lambda^2 + x^2}\, dx = |\lambda| \tag{14}$$

and

$$I_2 = \frac{1}{\pi} \int\limits_{-\infty}^{+\infty} \frac{\lambda x}{\lambda^2 + x^2}\, dx = 0 \ . \tag{15}$$

Since $|\lambda| = I_1 + iI_2$, we obtain

$$|\lambda| = \frac{1}{\pi} \int\limits_{-\infty}^{+\infty} \frac{\lambda^2 + i\lambda x}{\lambda^2 + x^2}\, dx = \frac{1}{\pi} \int\limits_{-\infty}^{+\infty} \left[1 - \frac{ix}{ix - \lambda} \right] dx \tag{16}$$

and thence

$$\sum_{j=1}^{n} |\lambda_j| = \sum_{j=1}^{n} \frac{1}{\pi} \int\limits_{-\infty}^{+\infty} \left[1 - \frac{ix}{ix - \lambda_j} \right] dx = \frac{1}{\pi} \int\limits_{-\infty}^{+\infty} \left[n - \sum_{j=1}^{n} \frac{ix}{ix - \lambda_j} \right] dx \ , \tag{17}$$

from which (12) follows immediately.   □

Since COULSON-type integral formulas will frequently be used in the following text and since they are rather important, we shall illustrate Eq. (12) by the example of fulvene, whose molecular graph is G_1

G_1

and whose characteristic polynomial $\varphi(G_1, x) = x^6 - 6x^4 + 8x^2 - 2x - 1$. Therefore, $\varphi'(G_1, x) = 6x^5 - 24x^3 + 16x - 2$ and formula (7.8) reads

$$E(G_1) = \frac{1}{\pi} \int\limits_{-\infty}^{+\infty} \left[6 - \frac{6x^6 + 24x^4 + 16x^2 + 2ix}{x^6 + 6x^4 + 8x^2 + 1 + 2ix} \right] dx \tag{18}$$

where both the numerator and the nominator are divided by $i^6 = -1$. It is by no means easy to calculate the above integral and even to see that its imaginary part is equal to zero. Hence the usage of Theorem 12.2 for numerical calculation of E would be a rather inappropriate idea.

Theorem 12.2, however, gives some insight into the dependence of $E(G)$ on the structure of G.

Let us first point at a few other integral formulas which are straightforward corollaries of Theorem 12.2.

Theorem 12.3. *Let G_1 and G_2 be graphs with equal number of vertices. Then*

$$E(G_1) - E(G_2) = \frac{1}{\pi} \int_{-\infty}^{+\infty} \ln |\varphi(G_1, ix)/\varphi(G_2, ix)| \, dx \, . \tag{19}$$

Proof. From Eq. (12),

$$E(G_1) - E(G_2) = -\frac{1}{\pi} \int_{-\infty}^{+\infty} x(d/dx) \ln \left[\varphi(G_1, ix)/\varphi(G_2, ix) \right] dx \, . \tag{20}$$

By partial integration we get

$$E(G_1) - E(G_2) = -\frac{x}{\pi} \ln \left[\varphi(G_1, ix)/\varphi(G_2, ix) \right]\Big|_{-\infty}^{+\infty}$$

$$+ \frac{1}{\pi} \int_{-\infty}^{+\infty} \ln \left[\varphi(G_1, ix)/\varphi(G_2, ix) \right] dx \, . \tag{21}$$

Equation (19) follows now from

$$x \ln \left[\varphi(G_1, ix)/\varphi(G_2, ix) \right] \to 0 \qquad \text{for } x \to \pm \infty \tag{22}$$

and from the fact that $E(G)$ is necessarily a real number. $\square$

Theorem 12.4. *For a graph G with n vertices,*

$$E(G) = \frac{1}{\pi} \int_{-\infty}^{+\infty} x^{-2} \ln |\psi(G, x)| \, dx \tag{23}$$

where

$$\psi(G, x) = (-ix)^n \, \varphi(G, i/x) \, . \tag{24}$$

Proof. Choose in Theorem 12.3 G_2 to be the null graph with n vertices. Then $\varphi(G_2, x) = x^n$ and $E(G_2) = 0$. Theorem 12.4 results now after an appropriate change of variables $x \to x^{-1}$. $\square$

Substituting (4.21) into (23) we obtain the explicit dependence of $E(G)$ on the coefficients of the characteristic polynomial:

$$E(G) = (2\pi)^{-1} \int_{-\infty}^{+\infty} x^{-2} \ln \left\{ \left[\sum_j (-1)^j a_{2j} x^{2j} \right]^2 + \left[\sum_j (-1)^j a_{2j+1} x^{2j+1} \right]^2 \right\} dx \, . \tag{25}$$

Substitution of the SACHS theorem (see paragraph 4.3.3) into Eq. (25) reveals the rather perplexing, but analytically well defined dependence of the total π-electron energy on the structure of the molecular graph. In principle, all questions concerning the dependence of $E(G)$ on molecular topology can be answered by combining (25) with the SACHS theorem. However, due to the rather complicated form of (25), such an analysis is by no means simple and many relevant problems in this field still remain unsolved.

In the case of alternant hydrocarbons formula (25) is considerably simplified because of Eq. (6.56). Thus, if G is a bipartite graph with $a + b$ vertices,

$$E(G) = \frac{1}{\pi} \int\limits_{-\infty}^{+\infty} x^{-2} \ln\left[\sum_{k=0}^{a} b(G, k)\, x^{2k} \right] dx . \tag{26}$$

If G is acyclic, then we have a further simplification:

$$E(G) = \frac{1}{\pi} \int\limits_{-\infty}^{+\infty} x^{-2} \ln\left[\sum_{k=0}^{[n/2]} m(G, k)\, x^{2k} \right] dx . \tag{27}$$

Equation (27) follows from (26) because of Eqs. (6.1) and (6.4).

We point now at three interesting consequences of (26) and (27).

Corollary 12.4.1. *If $G \in \mathscr{G}_{a,b}$, then $E(G)$ is a monotonically increasing function of the coefficients $b(G, k)$, $k = 1, 2, \ldots, a$. (These coefficients are, of course, defined via Eq. (6.56).)*

Corollary 12.4.2. *If $G \in \mathscr{T}_n$, then $E(G)$ is a monotonically increasing function of the matching numbers $m(G, k)$, $k = 1, 2, \ldots, [n/2]$. (These numbers are extensively discussed in paragraphs 4.2.2 and 6.1.3.)*

Corollary 12.4.3. *Let $G \in \mathscr{G}_{a,b}$ and let its characteristic polynomial be given by Eq. (6.56). Let ξ_k and τ_k be real numbers such that $0 \leq \xi_k \leq b(G, k)$ and $\tau_k \geq 0$, $k = 0, 1, \ldots, a$. Then*

$$\frac{1}{\pi} \int\limits_{-\infty}^{+\infty} x^{-2} \ln\left[\sum_{k=0}^{a} [b(G, k) - \xi_k]^2\, x^{2k} \right] dx \leq E(G)$$

$$\leq \frac{1}{\pi} \int\limits_{-\infty}^{+\infty} x^{-2} \ln\left[\sum_{k=0}^{a} [b(G, k) + \tau_k]^2\, x^{2k} \right] dx . \tag{28}$$

As a special case of the above inequalities we have a lower bound for $E(G)$ [102],

$$E(G) \geq J_1 + J_2 \tag{29}$$

where

$$J_1 = \frac{2}{\pi} \int_0^1 x^{-2} \ln \left[1 + b(G, 1) x^2 + b(G, 2) x^4 \right] dx \tag{30}$$

$$J_2 = \frac{2}{\pi} \int_1^\infty x^{-2} \ln \left[b(G, a-1) x^{2a-2} + b(G, a) x^{2a} \right] dx . \tag{31}$$

The estimate (29) is only one among the many possible lower bounds for E which can be deduced from Corollary 12.4.3. However, in this particular case J_1 and J_2 can be calculated by elementary methods of integration. Thus we have

$$J_1 = \frac{4}{\pi}(\alpha \arctan \alpha + \beta \arctan \beta) - \frac{2}{\pi} \ln \left[1 + b(G, 1) + b(G, 2) \right] \tag{32}$$

where

$$\alpha = \left\{ \frac{1}{2} \left[b(G, 1) + \sqrt{b(G, 1)^2 - 4b(G, 2)} \right] \right\}^{1/2} \tag{33}$$

$$\beta = \left\{ \frac{1}{2} \left[b(G, 1) - \sqrt{b(G, 1)^2 - 4b(G, 2)} \right] \right\}^{1/2} \tag{34}$$

and

$$J_2 = \frac{2}{\pi}\{n - 2 + \pi\gamma - 2\gamma \arctan \gamma + \ln \left[b(G, a-1) + b(G, a) \right]\} \tag{35}$$

where

$$\gamma = [b(G, a)/b(G, a-1)]^{1/2} . \tag{36}$$

It has been found that an unexpectedly good linear correlation exists between E and its lower bound $J_1 + J_2$ [102].

12.5 Some Further Applications of the Coulson Integral Formula

Suppose that we are able to show that for two bipartite graphs G_1 and G_2 the inequalities

$$b(G_1, k) \geq b(G_2, k) \tag{37}$$

are satisfied for all values of $k \geq 0$. Then from Corollary 12.4.1 it immediately follows that

$$E(G_1) \geq E(G_2) . \tag{38}$$

Equality in (38) is reached only if (37) is an equality for all k which in practice will occur only if G_1 and G_2 are cospectral. (Note that in (37) it is not required that the graphs G_1 and G_2 have equal number of vertices. Nevertheless, the case when G_1 and G_2 have unequal number of vertices is of little interest for us.)

If (37) holds for all k, we will write $G_1 > \cdot\, G_2$ or $G_2 \,\cdot< G_1$. If G_1 and G_2 are acyclic graphs, then the relation $G_1 > \cdot\, G_2$ reduces to the relation $G_1 > G_2$, discussed in detail in paragraph 6.1.4.

Hence, if G_1 and G_2 are acyclic graphs, then the relation $G_1 > G_2$ implies

$$E(G_1) \geqq E(G_2) \,. \tag{39}$$

As a consequence, each result given in Theorem 11.7 yields a pertinent inequality for total π-electron energy. We shall formulate here only some of them.

Theorem 12.5. *If $T \in \mathcal{T}_n$, then*

$$E(K_{1,\,n-1}) \leqq E(T) \leqq E(P_n) \tag{40}$$

with equality only if $T \simeq K_{1,\,n-1}$ or $T \simeq P_n$, respectively.

It should be mentioned that $E(K_{1,\,n-1}) = 2\sqrt{n-1}$ and $E(P_n) = 2\,\mathrm{cosec}\,[\pi/(2n+2)] - 2$ if n is even and $E(P_n) = 2\cotan\,[\pi/(2n+2)] - 2$ if n is odd.

Theorem 12.6. *If $P_n(j)$ is the tree defined in paragraph 11.2.4, then for $n = 4k - 1$ and $n = 4k$,*

$$E(P_n(1)) \geqq E(P_n(3)) \geqq E(P_n(5)) \geqq \ldots \geqq E(P_n(2k-1))$$
$$\geqq E(P_n(2k)) \geqq E(P_n(2k-2)) \geqq \ldots \geqq E(P_n(4)) \geqq E(P_n(2)) \tag{41}$$

whereas for $n = 4k + 1$ and $n = 4k + 2$,

$$E(P_n(1)) \geqq E(P_n(3)) \geqq E(P_n(5)) \geqq \ldots \geqq E(P_n(2k-1)) \geqq E(P_n(2k+1))$$
$$\geqq E(P_n(2k)) \geqq E(P_n(2k-2)) \geqq \ldots \geqq E(P_n(4)) \geqq E(P_n(2)) \,. \tag{42}$$

We show now a similar result which holds for bipartite graphs [183].

Theorem 12.7. *Let the graph $P_n(j, v)\, G$ be obtained by joining the vertex v of a bipartite graph G to the j-th vertex of the path P_n:*

$$
\begin{array}{c}
G \\
\big\downarrow v \\
\underset{1\;\;2}{\circ\!-\!\circ}-\cdots-\underset{j-1\;\;j\;\;j+1}{\circ\!-\!\circ\!-\!\circ}-\cdots-\underset{n}{\circ}
\end{array}
$$

$$P_n(j, v)\,G$$

Then all the inequalities given in Theorem 12.6 remain valid if $P_n(j)$ is replaced by $P_n(j, v) G$.

Proof. It is sufficient to demonstrate that

$$P_n(1, v) G > \cdot P_n(3, v) G > \cdot \ldots > \cdot P_n(4, v) G > \cdot P_n(2, v) G . \tag{43}$$

Notice that the edge joining the vertices j and v is a bridge. Hence we may apply Eq. (4.30), which results in

$$\varphi(P_n(j, v) G) = \varphi(P_n \cup G) - \varphi(P_{j-1} \cup P_{n-j} \cup G - v) \tag{44}$$

or in coefficient form

$$b(P_n(j, v) G, k) = b(P_n \cup G, k) + b(P_{j-1} \cup P_{n-j} \cup G - v, k - 1) . \tag{45}$$

Now, the first term on the right-hand side of the above identity is independent of j, whereas

$$b(P_{j-1} \cup P_{n-j} \cup G - v, k) = \sum_{l=0}^{k} b(P_{j-1} \cup P_{n-j}, l) \, b(G - v, k - l) . \tag{46}$$

Consequently $P_n(j, v) G > \cdot P_n(j^*, v) G$ if and only if $P_{j-1} \cup P_{n-j} > P_{j^*-1} \cup P_{n-j^*}$. The rest of the proof is now a simple application of Lemma 6.7. $\square$

Many other results similar to Theorems 12.5–12.7 are also known [100, 122].

In the subsequent chapter two topologically related isomers S and T will be considered and it will be shown (see Eq. (13.3)) that

$$\varphi(T) - \varphi(S) = [\varphi(A - k) - \varphi(A - l)]^2 \tag{47}$$

where A is a graph (from which the S-, T-isomers are constructed) and k and l are two of its vertices (see Fig. 13.2a).

Theorem 12.8 [92]. *If the S-, T-isomers are alternant hydrocarbons, then*

$$E(S) \geqq E(T) \tag{48}$$

with equality if and only if $\varphi(A - k) = \varphi(A - l)$.

Proof. The theorem requires that the graphs S and T and therefore also A are bipartite. Hence their characteristic polynomials obey

$$\varphi(S, ix) = i^{2n} \sum_j b(S, j) \, x^{2n-2j} \tag{49}$$

and

$$\varphi(T, ix) = i^{2n} \sum_j b(T, j) \, x^{2n-2j} \tag{50}$$

where n is the number of vertices of A and thus $2n$ is the number of vertices of S and T. In addition to this, $\varphi(A - k) - \varphi(A - l)$ is a polynomial of degree $n - 1 - 2\xi$ where ξ is some unspecified integer. Consequently

$$[\varphi(A - k, ix) - \varphi(A - l, ix)]^2 = i^{2(n-1-2\xi)}\left\{\sum_j [b(A - k, j) - b(A - l, j)] x^{n-2j}\right\}^2.$$

$$(51)$$

Combining the above three equations with (7.18) we easily conclude that

$$\varphi(S, ix)/\varphi(T, ix) = 1 - i^{4\xi+2}\left\{\sum_j [b(A - k, j) - b(A - l, j)] x^{n-2j}\right\}^2 \geqq 1 \qquad (52)$$

because $-i^{4\xi+2} = +1$.

Theorem 12.8 is obtained now by applying Theorem 12.3. □

It should be noted that Theorem 12.8 cannot be extended to non-alternant S-, T-isomers. Examples of non-alternant S-, T-isomeric pairs are known, for which $E(S) < E(T)$ [92].

As a third application of the COULSON integral formula we describe the following method for designing approximate topological expressions for $E(G)$ [99, 120].

It is clear that the full information on the dependence of $E(G)$ on the structure of G is contained in Eq. (25). We may write (25) as

$$E(G) = \frac{1}{\pi} \int_{-\infty}^{+\infty} F(x)\, dx \qquad (53)$$

where the form of $F(x)$ is seen from (25). The knowledge of the behaviour of $F(x)$ in the entire interval $(-\infty, +\infty)$ is required if one wants to predict $E(G)$ on the basis of (53), and this is exactly what is missing. On the other hand, some of the properties of $F(x)$ can be easily established by analytical methods, especially for large and for near-zero values of x:

1° $F(x) = -F(-x)$;
2° $F(x)$ has a unique maximum for $x = 0$ and $F(0) = m$;
3° $F(x)$ monotonically decreases in the interval $(0, \infty)$ and vanishes for $x \to \infty$; hence $F(x)$ is bell-shaped;
4° for large values of x, $F(x)$ behaves like $nx^{-2} \ln x$, provided $a_n \neq 0$.

Suppose now that $F^*(x)$ is another function having the properties 1°–4° and that the integral

$$E^*(G) = \frac{1}{\pi} \int_{-\infty}^{+\infty} F^*(x)\, dx \qquad (54)$$

can be explicitly calculated. Then it is reasonable to expect that $E^*(G)$ will be a satisfactory approximate formula for $E(G)$.

As a matter of fact, a great number of simple analytical functions with the desired properties could be constructed, leading to a whole class of approximate topological formulas for $E(G)$ [99, 120]. Moreover, some of these functions depend on one or even two unspecified parameters. Then the approximation can be further improved by adjusting a parameter so that $F^*(x_0) = F(x_0)$ for some selected value of x_0 (usually $x_0 = 1$).

For example,

$$F^*(x) = [m + nW \ln(1 + Bx^2)]/(1 + 2Wx^2) \tag{55}$$

where

$$W = [m - F(1)]/[2F(1) - n \ln(1 + B)] \tag{56}$$

and B is an arbitrary constant, $B > \sqrt[3]{20} - 1$, is a function behaving according to $1°$–$4°$ and leading to the approximate topological formula

$$E^* = (2W)^{-1/2} m + (2W)^{1/2} n \ln[(B/2W)^{1/2} + 1]. \tag{57}$$

12.6 Bounds for Total π-Electron Energy

There are nowadays very many known bounds for the total π-electron energy. The first and probably the most important of these inequalities was obtained by MCCLELLAND [144].

Theorem 12.9. *Let G be a graph with n vertices, m edges and adjacency matrix A. Then*

$$\sqrt{2m + n(n-1)|\det A|^{2/n}} \leqq E(G) \leqq \sqrt{2mn}. \tag{58}$$

Proof of the left inequality. We start with an identity for E based on Eq. (4.27):

$$E^2 = \left[\sum_{j=1}^{n} |\lambda_j|\right]^2 = \sum_{j=1}^{n} \lambda_j^2 + \sum_{j \neq k} |\lambda_j||\lambda_k| = 2m + \sum_{j \neq k} |\lambda_j||\lambda_k|. \tag{59}$$

Since for non-negative numbers the arithmetic mean is not smaller than the geometric mean,

$$\frac{1}{n(n-1)} \sum_{j \neq k} |\lambda_j||\lambda_k| \geqq \left[\prod_{j \neq k} |\lambda_j||\lambda_k|\right]^{1/n(n-1)}$$

$$= \left[\prod_{j=1}^{n} |\lambda_j|^{2(n-1)}\right]^{1/n(n-1)} = \left[\prod_{j=1}^{n} |\lambda_j|\right]^{2/n} = |\det A|^{2/n}. \tag{60}$$

Combination of the above two relations gives the lower bound for energy. $\qquad\square$

Proof for the right inequality. From the identity

$$\sum_{j<k} [\,|\lambda_j| - |\lambda_k|\,]^2 = n \sum_{j=1}^{n} |\lambda_j|^2 - \left[\sum_{j=1}^{n} |\lambda_j| \right]^2 \tag{61}$$

one gets $2nm - E^2 \geqq 0$. $\quad \square$

Corollary 12.9.1. *If* $\det A \neq 0$, *then* $E \geqq [2m + n(n-1)]^{1/2} \geqq n$.

Theorem 12.9 was later improved [98].

Theorem 12.10. *Let* $U = 2m - n|\det A|^{2/n}$. *Then for any graph* G

$$U \leqq 2nm - E^2 \leqq (n-1)\,U \tag{62}$$

while for bipartite graphs

$$2U \leqq 2nm - E^2 \leqq (n-2)\,U\,. \tag{63}$$

Note that $U \geqq 0$.
Another recent improvement of MᴄCʟᴇʟʟᴀɴᴅ's upper bound was proposed by
Tüʀᴋᴇʀ [177].

Theorem 12.11. *If* G *is a bipartite graph, then*

$$E(G) \leqq 2\sqrt{m + [n(n-2)\,b(G,2)/2]^{1/2}} \leqq \sqrt{2mn}\,. \tag{64}$$

Proof will be provided only for the first of the above two inequalities.
For the bipartite graphs, because of the pairing theorem (see paragraph 6.3.2),

$$b(G,2) = \sum_{j<k}^{n/2} \lambda_j^2 \lambda_k^2\,. \tag{65}$$

Using the fact that the mean value of squares is not less than the square of the
mean value, we attain

$$\left[\binom{n/2}{2}^{-1} \sum_{j<k}^{n/2} \lambda_j^2 \lambda_k^2 \right]^{1/2} \geqq \binom{n/2}{2}^{-1} \sum_{j<k}^{n/2} \lambda_j \lambda_k = \frac{1}{2} \binom{n/2}{2}^{-1} \left[\left(\sum_{j=1}^{n/2} \lambda_j \right)^2 - \sum_{j=1}^{n/2} \lambda_j^2 \right]$$

$$= \frac{1}{2} \binom{n/2}{2}^{-1} [(E/2)^2 - m] \tag{66}$$

which after necessary transformations results in the required upper bound. $\quad \square$

Some less elementary upper and lower bounds for total π-electron energy were
obtained in [111] and [121].

Theorem 12.12. *Let the number of 3- and 4-membered rings in the conjugated system under consideration be denoted by n_3 and n_4, respectively. If its molecular graph has n_+ positive and n_- negative eigenvalues, then*

$$12n_3\{n_+n_-/[(n_+ + n_-)\,Q - 4m^2]\}^{1/2} \leq E(G)$$

$$\leq [2m(n_+ + n_-) - (n_+ - n_-)^2\,(Q/2m - 9n_3^2/m^2]^{1/2} \tag{67}$$

where $Q = 8n_4 + 2D - 2m$ and D is the sum of the squares of the degrees of the vertices of G.

Note that in the chemically most common cases, $n_3 = 0$ and $n_+ = n_- = n/2$ and then the above inequalities reduce to $0 \leq E(G) \leq (2mn)^{1/2}$. Hence, Theorem 12.12, despite its complicated form, is essentially weaker than Theorem 12.11.

Theorem 12.13. *Consider the system of equations*

$$\alpha^2 + (n/2 - 1)\,\beta^2 = m \tag{68}$$

$$\alpha\beta^{n/2-1} = K, \qquad K > 0. \tag{69}$$

Let α_1, β_1 be their solution, such that $\alpha_1 > \beta_1 > 0$. Let α_2, β_2 be another solution of the above system, such that $\beta_2 > \alpha_2 > 0$. Let $E_{min} = 2\alpha_1 + (n - 2)\,\beta_1$ and $E_{max} = 2\alpha_2 + (n - 2)\,\beta_2$. Then for a benzenoid hydrocarbon with n carbon atoms, m carbon-carbon bonds and K KEKULÉ structures,

$$E_{min} \leq E < E_{max}. \tag{70}$$

It has been shown that [121]

$$E_{min} \approx 2\xi + 2\tau(K/\xi)^{1/\tau} \tag{71}$$

and

$$E_{max} \approx 2(m\tau)^{1/2} + 2K(\tau/m)^{\tau/2} \tag{72}$$

where

$$\xi = \{m - K\tau[n - \tau(K^2/m)^{1/\tau}]^{-1/2}\}^{1/2} \tag{73}$$

and

$$\tau = n/2 - 1. \tag{74}$$

12.7 More on the McClelland Formula

The expression $(2mn)^{1/2}$ occurs in numerous approximate formulas and bounds for total π-electron energy. Already in [144] it was observed that the expression $a(2mn)^{1/2}$, where a is an empirically determined constant, is capable of reproducing $E(G)$ fairly well.

The following two results will give insight into this point from a somewhat different angle.

In Sect. 4.3 we have shown that the graph eigenvalues conform to the Eqs. (4.24) and (4.27). These can be understood as certain information about the distribution of the graph eigenvalues. On the other hand, according to (4), E is a function of the graph eigenvalues and must obviously be sensitive to their distribution. Thus we may pose the question: Which is the maximum possible value of E, Eq. (4) if the graph eigenvalues are constrained to the conditions (4.24) and (4.27) and only to these conditions?

Assuming that n is even and that the occupation numbers of the MO's satisfy the conditions $\eta_1 = \eta_2 = ... = \eta_{n/2} = 2$ and $\eta_{n/2+1} = ... = \eta_n = 0$, one obtains the following result [108].

Theorem 12.14. *The eigenvalue distribution which gives a maximum value for E and which simultaneously fulfils the relations (4.24) and (4.27) is*

$$\lambda_1 = \lambda_2 = ... = \lambda_{n/2} = (2m/n)^{1/2} \tag{75}$$

$$\lambda_{n/2+1} = ... = \lambda_n = -(2m/n)^{1/2} . \tag{76}$$

The maximum value for E is then $(2mn)^{1/2}$.

Theorem 12.14 implies, of course, MᴄCʟᴇʟʟᴀɴᴅ's upper bound. In fact, Theorem 12.14 in a certain sense explains the origin of his estimate. It is an interesting and not fully obvious finding that $(2mn)^{1/2}$ becomes equal to $E(G)$ only under the assumption that all bonding MO's (as well as all antibonding MO's) have equal energies.

However, if we consider a completely different hypothesis, namely that the MO energy levels are uniformly distributed (and hence no two of them degenerate):

$$\lambda_j - \lambda_{j+1} = \Delta \tag{77}$$

with Δ being a constant, independent of j, $j = 1, 2, ... , n - 1$, then a direct calculation gives

$$E(G) = f(n) \, (2mn)^{1/2} \tag{78}$$

where

$$f(n) = (n/2)[3/(n^2 - 1)]^{1/2} . \tag{79}$$

The function $f(n)$ rapidly converges to a constant value $f(n) = 0.87$ with increasing n. This limit is close to $a = 0.92$ determined empirically [144]. Therefore, one may conclude that MᴄCʟᴇʟʟᴀɴᴅ's approximate formula $a(2mn)^{1/2}$ is in a certain sense related to, and a consequence of the (tacit) assumption of a uniform distribution of the molecular orbital energy levels.

The apparent disadvantage of any approximate topological formula for total π-electron energy of the form $E^* = E^*(m, n)$ is that it cannot distinguish between isomers, i.e. predicts equal total π-electron energy for all isomers [115]. A considerable variety of approximate formulas has been proposed in the last 20–30 years in order to understand the factors causing energy differences between isomers[1]. These formulas, however, are often much less successful than their complicated algebraic form may suggest.

A systematic study [117] of all known approximate formulas for total π-electron energy depending on three (or less) invariants of the molecular graph–number of vertices, number of edges and number of KEKULÉ structures (perfect matchings) gave the following surprising result. Among 7 formulas containing one empirical parameter, the best was $0.908\,(2mn)^{1/2}$. Among 9 formulas containing two empirical parameters, the best was $0.899\,(2mn)^{1/2} + 0.426$. Among formulas containing three empirical parameters the best found was $0.714n + 0.566m + 0.395$. All 13 examined formulas depending on the number of KEKULÉ structures (which distinguish isomers) were found to be inferior to the above mentioned ones.

12.8 Conclusion: Factors Determining the Total π-Electron Energy

The results outlined in the preceding section strongly suggest that m and n are the dominant factors determining E. If any simple analytical expression can be used to describe this dependence, then it is certainly the McCLELLAND formula.

The role of other graph invariants is much more obscure. It is known that $E(G)$ decreases with the increase of branching of the molecular skeleton [134]. The earlier belief that in a series of structurally similar isomers $E(G)$ increases linearly with the logarithm of $|\det A|$ had recently to be revised. HALL has namely shown that for benzenoid hydrocarbons, E is a linear function of K, the number of KEKULÉ structures (recall Eq. (6.70)) [124]. HALL's empirical formula

$$E = 0.442n + 0.788m + 0.34K(0.632)^{m-n} \tag{80}$$

having a quite unusual algebraic form, could be supported by graph-theoretical arguments [118].

Cyclic conjugation is another factor important for the complete understanding of the behaviour of total π-electron energy. It is related to the presence of cycles in the molecular graph. Details of the theory of cyclic conjugation are reviewed elsewhere [110, 112, 119].

12.9 Use of Total π-Electron Energy in Chemistry

As already mentioned in Sect. 12.1, HMO total π-electron energies can be used to determine the heats of atomization of conjugated hydrocarbons. We now present this matter in more detail.

---

[1] For bibliography covering data until the middle of the 70's see [26] and [103]; more recent references can be found in [117, 120, 134].

The total electron energy of a conjugated molecule may be assumed to be equal to the sum of E_π, the total π-electron energy, and E_σ, the energy of the σ electrons. This latter energy can be further approximated as [173]

$$E_\sigma = n_C E_C^\sigma + n_H E_H^\sigma + \sum_i E_{CH}^\sigma(i) + \sum_j E_{CC}^\sigma(j) \, . \tag{81}$$

Here n_C and n_H are the number of carbon and hydrogen atoms, respectively; E_C^σ and E_H^σ are the σ-energies of isolated carbon and hydrogen atoms, respectively; $E_{CH}^\sigma(i)$ is the σ-electron energy of the i-th CH bond, $E_{CC}^\sigma(j)$ is the σ-electron energy of the j-th CC bond and the two summations on the right-hand side of (81) go over all chemical bonds in the molecule considered.

The first two terms in (81) will cancel in computations of heats of atomization or binding energies and need be examined no further. Adopting the usual assumption that all CH bonds have equal energy [90, 173], the third term in (81) will be reduced to $n_{CH} E_{CH}^\sigma$, with n_{CH} denoting the number of CH bonds.

It is evident that E_{CC}^σ depends somehow on the length of the corresponding CC bond. On the other hand, the CC bond lengths in conjugated hydrocarbons are to a great extent determined with the π-electron bond orders. Elaborating these arguments, one arrives at the relation [90]

$$E_{CC}^\sigma(j) + E_{CC}^\pi(j) = E_{CC}^0 + 2\beta p(j) \tag{82}$$

where E_{CC}^0 is a constant and $p(j)$ denotes the order of the j-th π-bond.

Such a consideration results finally in the following expression for the heat of atomization

$$\Delta H_a = - [n_{CH} E_{CH}^\sigma + n_{CC} E_{CC}^0 + \beta E] \tag{83}$$

where E is the total π-electron energy, given by Eqs. (5) and (6). The constants E_{CH}^σ, E_{CC}^0 and β are to be treated as semi-empirical parameters. A least-squares fitting, based on experimental heats of atomization gave [90]: $E_{CH}^\sigma = -411.09 \, kJ/mol$, $E_{CC}^0 = -325.18 \, kJ/mol$ and $\beta = -137.00 \, kJ/mol$.

The accuracy of formula (83) is, in fact, much better than could be concluded from Fig. 12.1. The agreement between experimental heats of atomization, those calculated by a PARISER-PARR-POPLE-type SCF MO method [83] and those calculated by Eq. (83) can be seen from the examples collected in Table 12.1.

Estimation of resonance energies is another field where HMO total π-electron energies are used in chemistry. As a matter of fact, a great number of different resonance energy concepts has been proposed in the last 10–20 years. It is not the intention of the present book to consider details of these resonance energies, but rather to mention a few of them.

Resonance energy is usually conceived as the difference between the total electron energy of a conjugated molecule and the energy of something that is called "reference structure". Since the reference structure is an imaginary moiety, it is not clear what should be considered as its energy. In several approaches the energy of the reference structure is obtained by summing certain bond-energy increments.

Table 12.1. Heats of atomization of selected conjugated hydrocarbons. All values are in *MJ/mol*

Molecule	Heats of atomization		
	Experimental	SCF MO [83]	Eq. (83)
Butadiene	4.057	4.057	4.054
Benzene	5.515	5.515	5.513
Hexatriene	5.895	5.867	5.872
Pentalene	—	6.825	6.805
Azulene	8.606	8.632	8.696
Naphthalene	8.743	8.743	8.741
Acenaphthylene	10.065	10.117	10.118
Anthracene	11.957	11.954	11.959
Phenanthrene	11.984	11.985	11.978
Perylene	16.599	16.610	16.607

For the reader's convenience four different sets of parameters are collected in Table 12.2. They have been proposed for the calculation of resonance energy by means of the formula (84):

$$RE = E - \sum_{bonds} E(bond) \tag{84}$$

Table 12.2. Parametrization schemes for the calculation of resonance energy via Eq. (84). All bond-energy parameters are in β-units. Note that in the variant of HESS and SCHAAD the value of the resonance energy depends on the KEKULÉ structure upon which the calculation is based; an averaging over all KEKULÉ structures is therefore necessary. In the method of JIANG, TANG and HOFFMANN it is not necessary to distinguish between single and double bonds

Variant	Bond type	Bond energy
Classical resonance energy	$C=C$	2.0000
(see e.g. [62])	$C-C$	0.0000
DEWAR resonance energy	$C=C$	2.0000
[83, 146]	$C-C$	0.5200
HESS-SCHAAD resonance energy	$H_2C=CH$	2.0000
[127, 128]	$HC=CH$	2.0699
	$H_2C=C$	2.0000
	$HC=C$	2.1083
	$C=C$	2.1716
	$HC-CH$	0.4660
	$HC-C$	0.4362
	$C-C$	0.4358
Parametrization by JIANG,	H_2C-CH	1.5898
TANG and HOFFMANN [134]	H_2C-C	1.4145
	$HC-CH$	1.2691
	$HC-C$	1.1328
	$C-C$	1.0211

where E is the HMO total π-electron energy, $E(bond)$ is the energy corresponding to a particular bond type and the summation goes over all CC bonds.

We also mention in passing the so called "topological resonance energy" [69, 116] in which the reference energy is calculated in a somewhat different manner. Here, namely, the reference energy is obtained from a formula fully analogous to Eqs. (5) and (6), where instead of graph eigenvalues λ_j one uses the zeros of the matching polynomial.

Resonance energies are usually quoted in discussions concerning the aromaticity of conjugated molecules. It is not the authors' intention to consider this matter. We rather refer the interested reader to the books [5, 65]. A general impression about resonance energies can be gained from Fig. 12.2.

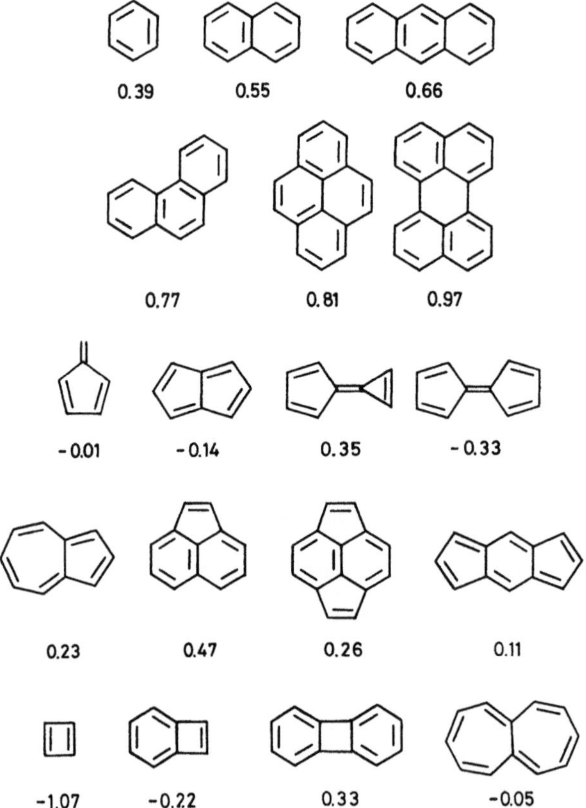

Fig. 12.2. Hess-Schaad resonance energies (in β units) of selected conjugated hydrocarbons. Observe that stable and easily obtainable compounds have large positive resonance energies; negative or near-zero resonance energies are associated with highly reactive conjugated species

Chapter 13

Topological Effect on Molecular Orbitals

Chemical experience is gained from the study of real existing molecules, which are thought to be composed of distinct particles of matter, namely atomic nuclei and electrons. Abstracting the molecular structure as far as possible one arrives at molecular topology. In the course of abstraction one and the same molecular topology may be obtained from very different molecules (see Fig. 2.1). Thus the pronounced differences in the chemical properties should be an exclusive consequence of the properties of the atoms used in the various realizations of a given topology. In view of this, one may wonder whether topology plays any role at all. The successful application of topological indices in various correlations (see Chap. 11) may be evaluated as some positive evidence, but not as a rigorously proved answer: neither the physical meaning of the topological indices nor the physical interrelations causing the correlations are sufficiently understood. Contrary to that, the topological effect on molecular orbitals (TEMO) [162] provides solid evidence that topology determines at least a frame within which series of physically and chemically diverse species may be realized.

13.1 Topologically Related Isomers

It has been shown in Chap. 2 that to each molecule a defined topological space is associated, which depends uniquely on the constitution of the molecule.

Definition 1: Two isomers with different constitutions are called *topologically related* if their respective topological spaces may be divided into two or more subspaces, such that they are pairwise isomorphic. Topologically related isomers are termed *topomers.*

Thus, the topological spaces of topomers differ only with respect to the connection of the respective subspaces. In Fig. 13.1 examples of topomers are given. The pairs of topomers are denoted by Roman numerals whereas the individual species of a given pair are distinguished by S and T. The topomers I are constructed from equal subunits while in the topomers II to V unequal subunits have been used. Note that the pairs I and II have the topomer T in common. If one supposes that the T topomer of a given pair is made from its S topomer, then some bonds must be first broken and then formed again, as indicated by the dashed lines in Fig. 13.1. The number of these bonds is 2 for I, II and III, 3 for IV and 4 for V. These bonds correspond to the adjacency relations which distinguish the topomer S from the topomer T. Hence,

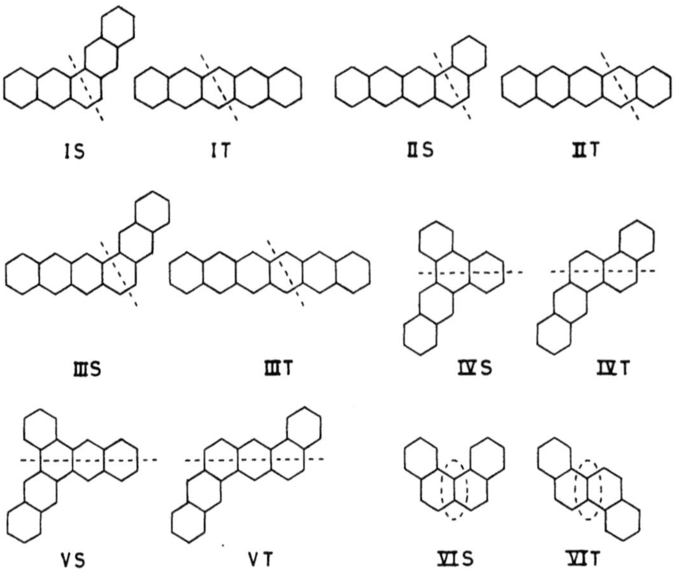

Fig. 13.1. Examples of topomeric aromatic hydrocarbons

the number of these bonds also indicates the number of reorganized adjacency relations. Writing down the ball-neighbourhoods (as defined and exemplified in Sect. 2.3),
one immediately sees that the effect of reorganized adjacency relations increases
relatively if the subunits are made unequal as well as if the number of reorganized
adjacency relations is increased. Consequently, the effect is greater in II and III than
in I, etc.

The pair VI can be understood as composed of two unequal moieties derived
from naphthalene and styrene, respectively. It may also be considered as formed
of three moieties: two terminal ones derived from styrene and a central moiety derived
from ethylene.

The manner in which a pair of topomers may be constructed is called a *topological
model*; it is characterized firstly by the number of subunits used and secondly by the
specifications by which these subunits are connected. The model 1 [162] used in the
construction of the pairs I to V works with two subunits; it is modified by the number
of bonds, $l = 2, 3, 4$ as described above. For the construction of the pair VI another
model, model 2, [95] is used which works with three subunits. Further models are
described elsewhere [92, 95, 155, 156, 157, 159, 162].

It should be noted that the connection of two univalent subunits by one bond
cannot generate a pair of topomers. Therefore the valencies of the subunits and the
number of connecting bonds must be at least 2. But even from two bivalent subunits
no pair of topomers can be constructed unless the sites of valency in the subunits
are non-equivalent. Thus in the case of two connecting bonds, $l = 2$, the models 1
and 2 used in the construction of the pairs I–III and VI, respectively, can be schematiz-
ed as shown in Fig. 13.2. Therein the vertices $k, l \in \mathscr{V}(A)$ as well as $p, q \in \mathscr{V}(B)$ are
non-equivalent.

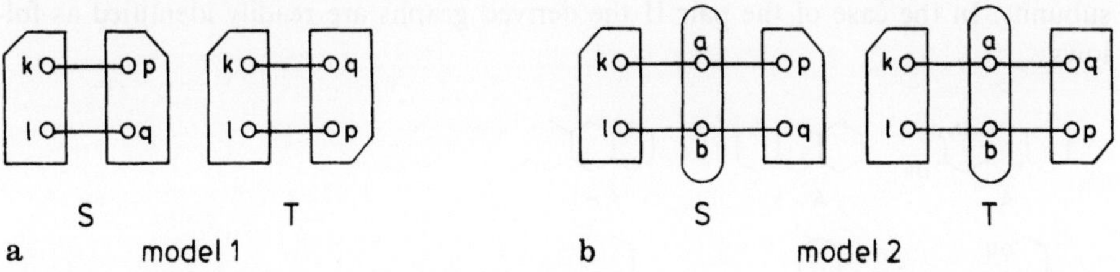

Fig. 13.2. Schematization of two topological models in the case of two connecting bonds (edges). The subunit containing the vertices k and l is denoted by A; the subunit containing p and q is B

Supposing that the subunits A and B are isomorphic, $A \simeq B$, some symmetry is induced in the topomers constructed in accordance with the model 1. The same is true for topomers according to model 2, provided the central moiety does not lower that symmetry. For instance, if model 1 is applied to the generation of planar topomers, then S and T exhibit at least $\mathbf{C}_{2v}$ and $\mathbf{C}_{2h}$ symmetry, respectively.

13.2 Interlacing Rule

Applying Eq. (4.29) to S and T of the model 1, $l = 2$, as depicted in Fig. 13.2a, one obtains for the respective characteristic polynomials,

$$\varphi(S, x) = \varphi(A, x)\,\varphi(B, x) - \varphi(A - k, x)\,\varphi(B - p, x) - \varphi(A - l, x)\,\varphi(B - q, x)$$

$$+ \varphi(A - k - l, x)\,\varphi(B - p - q, x)$$

$$- 2[\Sigma\,\varphi(A - P_{kl}, x)]\,[\Sigma\,\varphi(B - P_{pq}, x)]$$

$$\varphi(T, x) = \varphi(A, x)\,\varphi(B, x) - \varphi(A - k, x)\,\varphi(B - q, x)$$

$$- \varphi(A - l, x)\,\varphi(B - p, x) + \varphi(A - k - l, x)\,\varphi(B - p - q, x)$$

$$- 2[\Sigma\,\varphi(A - P_{kl}, x)]\,[\Sigma\,\varphi(B - P_{pq}, x)]$$

$$(1)$$

where P_{kl} and P_{pq} denote elementary paths (see paragraph 4.1.4), connecting the vertices k and l, and p and q, respectively. The summations in (1) go over all paths which connect the respective vertices.

The difference of the two polynomials given in (1) is denoted by $\Delta(x)$ and direct calculation shows that

$$\Delta(x) = \varphi(T, x) - \varphi(S, x)$$

$$= [\varphi(A - k, x) - \varphi(A - l, x)]\,[\varphi(B - p, x) - \varphi(B - q, x)] . \qquad (2)$$

This expression may be applied to the difference of the characteristic polynomials of S and T in the pairs I to III of Fig. 13.1, but not to the other pairs because in IV and V there are more than 2 connecting bonds whereas VI is constructed from three subunits. In the case of the pair II the derived graphs are readily identified as follows:

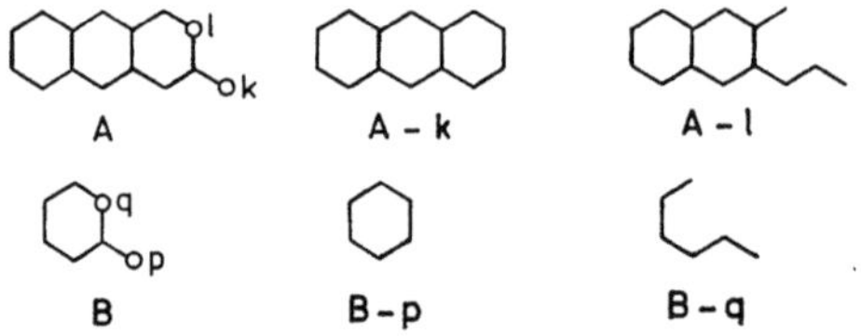

The vertices $k, l \in \mathscr{V}(A)$ and $p, q \in \mathscr{V}(B)$ must be non-equivalent in order to achieve that S and T represent different chemical species. Eq. (2) illuminates this from another point of view: If the vertices of one of these pairs were equivalent, then $\Delta(x)$ would be identically equal to zero and, consequently, $\varphi(T, x) = \varphi(S, x)$.

The vertices corresponding to $k, l \in \mathscr{V}(A)$ and $p, q \in \mathscr{V}(B)$ in the examples given in Fig. 13.1 are evidently non-equivalent since they cannot be mapped automorphically onto each other. Thus the given examples exhibit *constitutional (structural) isomerism*. Notice that $\Delta(x)$ will not be identically equal to zero even if the vertices k and l are topologically equivalent, but realized in the molecular structure by atoms of different kind [163]. Examples constructed in that way exhibit *positional isomerism*, as for instance:

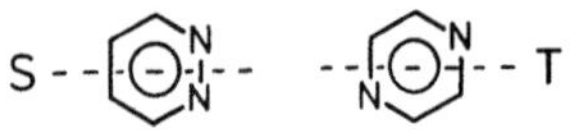

In the case that the subunits are isomorphic, as realized in the pair I, $A \simeq B$, $A - k \simeq B - p$ and $A - l \simeq B - q$, the respective polynomials are equal and, as a consequence of this, the bilinear expression (2) reduces to a perfect square

$$\Delta(x) = [\varphi(A - k, x) - \varphi(A - l, x)]^2 \geq 0 . \tag{3}$$

In this case S denotes the topomer in which the equivalent centers of A and B are connected pairwise by bonds (edges). Obviously, in this case $\Delta(x)$ is non-negative in the complete range of the variable x and the following inequality holds:

$$\varphi(T, x) \geq \varphi(S, x); \quad x \in (-\infty, +\infty) . \tag{4}$$

A consequence of (4) was already presented in Theorem 12.8. Another consequence of (4) is the following important result.

Theorem 13.1. *Let S and T be a pair of topomers made from isomorphic subunits according to Fig. 13.2a and let $\lambda_j(S)$ and $\lambda_j(T)$ denote the zeros of the characteristic polynomials of S and T, respectively. Then,*

$$\lambda_1(S) \geq \lambda_1(T) \geq \lambda_2(T) \geq \lambda_2(S) \geq \ldots \geq \lambda_{2j-1}(S) \geq \lambda_{2j-1}(T)$$

$$\geq \lambda_{2j}(T) \geq \lambda_{2j}(S) \geq \ldots . \tag{5}$$

The genesis of (5) is schematically illustrated in Fig. 13.3. The pattern of eigen-values according to (5) has been termed *topological effect on molecular orbitals* (TEMO) [162]. It is schematically shown in Fig. 13.4. Two independent proofs of Theorem 13.1 are given in [93, 114].

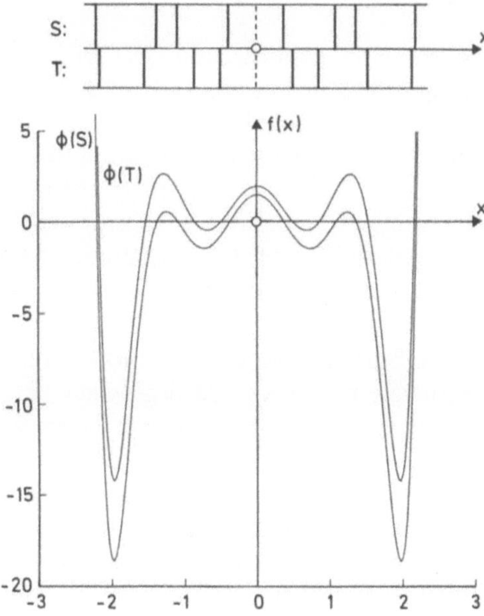

Fig. 13.3. Illustration of Theorem 13.1 and its relation to Eq. (4)

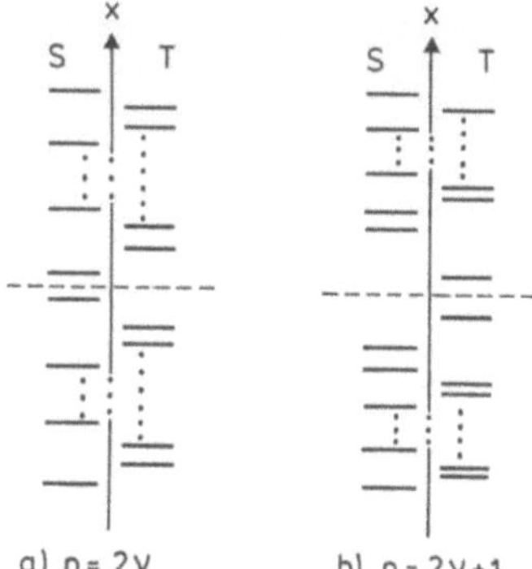

Fig. 13.4. Eigenvalue pattern of S and T

The TEMO pattern shown in Fig. 13.4 is obtained purely from the topology. Suppose that S and T are fully conjugated hydrocarbons. Then the pattern given in Fig. 13.4 must be identified with the Hückel MO energy levels of the topomers. Let n be the number of π-electrons of the subunit A. Then the MO patterns of S and T consist of $2n$ MO levels, n of which are doubly occupied. From (5) and Fig. 13.4 one arrives at the following conclusion.

Rule 1: If the number n of π-electrons of the building subunit A is even, then the HOMO-LUMO separation ΔE as well as the first ionisation potential IP_1 are smaller for S than for T, provided the HOMO level is not degenerate. If n is odd, the reverse is true.

This rule is in excellent agreement with experimental results as proved by the p-absorption bands and first ionization potentials of polycyclic aromatic hydrocarbons [12, 162].

As another more general consequence of (5), the photoelectron (PE) spectra of topomers should exhibit a similar interlacing as depicted in Fig. 13.4. In Sect. 13.3 it is shown that this pretentious demand is also well satisfied.

The isomorphism of the subunits A and B is a necessary presupposition for the formulation of (3)–(5), Theorem 13.1 and Rule 1. Let us now return to the more general case $A \neq B$. Then for an arbitrary value of x no direct conclusion can be drawn from Eq. (2) cocerning the sign of $\Delta(x)$ and the relative magnitude of the characteristic polynomials $\varphi(S, x)$ and $\varphi(T, x)$. Nevertheless, some rules may be extracted even in this case.

The difference function $\Delta(x)$ changes sign if the variable x passes through a real root of $\Delta(x) = 0$ which is either non-degenerate or has an odd degeneracy; it is readily seen that complex roots or double roots, fourfold roots etc. do not change the sign of $\Delta(x)$. Let $x_j, j = 1, 2, 3, ...$, denote the real roots of odd degeneracy of $\Delta(x) = 0$ and let them be labeled in decreasing order, $x_1 > x_2 > x_3 > ...$ Then the roots x_j determine open intervals, e.g. (x_{j+1}, x_j), in which either $\Delta(x) \geq 0$ or $\Delta(x) \leq 0$ holds. $\Delta(x)$ takes the value zero within such an interval if and only if the interval contains a real root of even degeneracy. It is readily seen that $\Delta(x)$ has the sign of $(-1)^j$ within the open interval (x_{j+1}, x_j) and opposite sign within the two adjacent intervals, provided $\Delta(x)$ is non-negative in the interval (x_1, ∞). Then in all intervals (x_{2k+1}, x_{2k}) there will be $\Delta(x) \geq 0$ and, hence, $\varphi(T, x) \geq \varphi(S, x)$. As said above, under these conditions (5) holds. On the other hand, within all intervals (x_{2k}, x_{2k-1}) there will be $\Delta(x) \leq 0$ and, consequently, $\varphi(T, x) \leq \varphi(S, x)$. This produces an *inversion* of the order of MO's and therefore within these intervals

$$... \geq \lambda_{2j-1}(T) \geq \lambda_{2j-1}(S) \geq \lambda_{2j}(S) \geq \lambda_{2j}(T) \geq ... . \tag{6}$$

Since the real roots x_j of odd degeneracy cause these inversions, they are termed *inversion points*.

The above considerations have shown that the MO spectra of pairs of topomers constructed according to the model 1 (Fig. 13.2a) exhibit an interlacing as expressed by the relations (5) and (6). If the subunits A and B are isomorphic, no inversion occurs, but inversions may appear if A and B are non-isomorphic.

The complexity of the expression for $\Delta(x)$ increases rapidly with the increase of the number l of connecting bonds. In the case of topomers made of two trivalent subunits, as realized by the pair IV, one has the following schematic model:

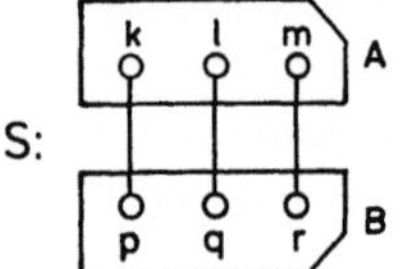

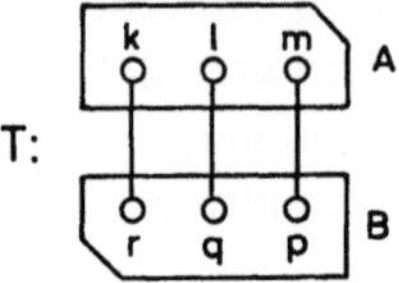

from which one derived

$$
\begin{aligned}
\Delta(x) = {} & [\varphi(A - k, x) - \varphi(A - m, x)]\,[\varphi(B - p, x) - \varphi(B - r, x)] \\
& - [\varphi(A - k - l, x) - \varphi(A - l - m, x)]\,[\varphi(B - p - q, x) - \varphi(B - q - r, x)] \\
& + 2[\Sigma\,\varphi(A - P_{kl}, x) - \Sigma\,\varphi(A - P_{lm}, x)]\,[\Sigma\,\varphi(B - P_{pq}, x) \\
& - \Sigma\,\varphi(B - P_{qr}, x)] \\
& - 2[\Sigma\,\varphi(A - m - P_{kl}, x) - \Sigma\,\varphi(A - k - P_{lm}, x)] \\
& \times [\Sigma\,\varphi(B - r - P_{pq}, x) - \Sigma\,\varphi(B - p - P_{qr}, x)]\,.
\end{aligned}
\tag{7}
$$

Even if $A \simeq B$ this expression results in a sum of four squares with pairwise opposite signs. In the case of $l = 4$, the function $\Delta(x)$ has 32 bilinear terms [156]. Thus, for pairs of topomers in which the subunits are connected by more than two bonds the appearance of inversions cannot be excluded.

For illustration, in Table 13.1 the HMO eigenvalues of triphenylene (VII S) and chrysene (VII T) are given [15]. The respective characteristic polynomials are:

Table 13.1. Bonding HMO eigenvalues [15] of triphenylene (VII S) and chrysene (VII T) and inversion points. The antibonding eigenvalues are symmetrical to the bonding ones with respect to $x = 0$

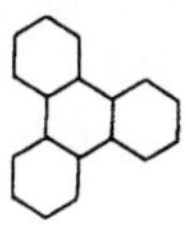

VII S	inversion points	VII T
2.532 089		
		2.499 046
		2.166 518
1.969 616		
1.969 616		
		1.700 759
		1.539 774
1.347 297		
------	1.285 778[a]	------
		1.285 774
1.285 575		
1.285 575		
		1.216 441
------	1.000 000	------
0.879 386		
		0.875 324
		0.792 335
0.684 041		
0.684 041		
		0.520 139

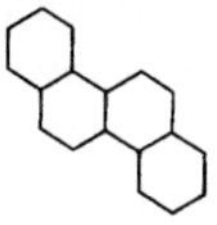

$$\varphi(\text{VII } S, x) = x^{18} - 21x^{16} + 180x^{14} - 825x^{12} + 223x^{10} - 3645x^8$$
$$+ 3627x^6 - 2106x^4 + 648x^2 - 81 \tag{8}$$

$$\varphi(\text{VII } T, x) = x^{18} - 21x^{16} + 180x^{14} - 824x^{12} + 2214x^{10} - 3605x^8$$
$$+ 3533x^6 - 1990x^4 + 577x^2 - 64 . \tag{9}$$

Thus the difference polynomial $\Delta(x)$ takes the form

$$\Delta(x) = x^{12} - 9x^{10} + 40x^8 - 94x^6 + 116x^4 - 71x^2 + 17 . \tag{10}$$

The real roots of $\Delta(x) = 0$ are $x_{1,4} = \pm 1.286$ and $x_{2,3} = \pm 1.0$. Thus, there are two intervals, (1.0, 1.286) and $(-1.286, -1.0)$ within which the MO pattern is inverted (see Table 13.1).

For the model 2 of Fig. 13.2b one obtains [92, 95, 155, 156]:

$$\Delta(x) = [\varphi(A - k, x) - \varphi(A - l, x)] [\varphi(B - p, x) - \varphi(B - q, x)] \varphi(C - a - b, x) . \tag{11}$$

Once again, if $A \simeq B$ and C has a symmetric structure such that $\varphi(C - a - b, x) = \gamma(x)^2$, then Eq. (11) takes the form of a square [156]:

$$\Delta(x) = [\varphi(A - k, x) - \varphi(A - l, x)]^2 \gamma(x)^2 \geqq 0 \tag{12}$$

and therefore the interlacing of the MO eigenvalues according to (5) has to be expected. Further, for this model the appearance of inversions can be excluded.

Model 2 may also be extended to tri-, tetra-, etc. valent subunits [156], but then the appearance of inversions cannot be excluded.

The topomers VI are considered to be constructed from three subunits as indicated in Fig. 13.1. They could also be thought of as formed according to model 1 from two non-equivalent subunits derived from naphthalene and styrene, respectively. In the second case due to $A \neq B$ inversions could not be excluded, but this can be done when model 2 is applied. This illustrates the usefulness of having different models at one's disposal.

13.3 PE Spectra of Topomers

The photoelectron (PE) spectrum of a molecule consists of a series of vertical ionisation potentials which are related to MO energies of the molecule, provided the KOOPMANS theorem [137] holds. Hence the PE spectrum of a molecule represents the upper part of its MO energy diagram. Thus the inspection of the PE spectra of topomers is the most rigorous proof of the physical relevance of TEMO because it allows an estimate of the extent to which the purely topological TEMO rule dominates in physical reality.

For such an estimate not the accurate location of the MO levels, but rather the satisfaction of the interlacing rule, which is the intrinsic characteristic of TEMO, has crucial significance. Having in mind that all material and geometrical features of the molecular structure considered are completely ignored in the topological treatment, some violations of TEMO should even be anticipated. Such violations must be understood as physically induced inversions. Since the TEMO rule is a consequence of Eq. (2), the absolute value of $\Delta(x)$ should indicate somehow the strength of the topological features when they compete with the physical ones in modeling the MO pattern as finally observed by means of PE spectra.

Table 13.2. The PE spectra of the topomers I–VI. All energies are in eV. Inversion points are indicated by dashed lines

IS / IT, Ref. [175]

IS	IT
	6.61
7.27	
7.39	
	7.92
	8.32
8.54	
8.90	
	9.01
	9.39
9.53	
9.66	
	9.80
	10.23
10.3	
10.5	

IIS / IIT, Ref. [175]

IIS	IIT
	6.61
6.97	
– – –	– – –
	7.92
8.00	
– – –	– – –
	8.32
8.45	
– – –	– – –
	9.01
9.06	
– – –	– – –
	9.39
9.50	
– – –	– – –
	9.80
10.00	
– – –	– – –
	10.23
10.31	
10.8	

IIIS / IIIT, Ref. [76]

IIIS	IIIT
	6.44
7.02	
7.47	
	7.55
	8.14
8.31	
– – –	– – –
	8.56
8.60	
9.16	
	9.36
	9.36
9.49	
9.90	
	9.95
	9.95
10.25	
– – –	– – –
	10.30

IVS / IVT, Ref. [175]

IVS	IVT
	7.20
7.39	
– – –	– – –
7.89	
	7.95
	8.24
8.28	
– – –	– – –
	9.05
9.14	
	9.25
9.39	
– – –	– – –
	9.91
9.92	
9.92	
	10.3

VS / VT, Ref. [175]

VS	VT
	7.14
7.30	
– – –	– – –
	7.4
7.48	
7.96	
	8.08
	8.67
8.72	
– – –	– – –
	8.96
9.14	
9.37	
	9.47
– – –	– – –
9.69	
	9.85
– – –	– – –
9.96	
	10.18

VIS / VIT, Ref. [175]

VIS	VIT
	7.59
7.60	
8.02	
	8.10
	8.68
8.98	
9.18	
	9.43
	9.72
9.96	
10.22	
	10.52

VIIS / VIIT, Ref. [175]

VIIS	VIIT
	7.59
7.88	
7.88	
	8.10
– – –	– – –
8.65	
	8.68
	9.43
9.68	
9.68	
	9.72
– – –	– – –
10.06	
	10.52

Model / Inversion summary

	IS IT	IIS IIT	IIIS IIIT	IVS IVT	VS VT	VIS VIT	VIIS VIIT
Model 1	1 / 2	1 / 2	1 / 2	1 / 3	1 / 4	2 / 2	1 / 3
	$A \simeq B$	$A \neq B$	$A \neq B$	$A \neq B$	$A \neq B$	$A \simeq B$	$A \simeq B$
Inv.:	0	6	2	3	4	0	2

In Table 13.2 the PE spectra of the topomers I to VI are given. All examples are taken from the class of polycyclic aromatic hydrocarbons because the PE spectra of these compounds usually have an exceptionally high number of well-resolved peaks. For TEMO within other classes of compounds see [88, 92, 95, 148, 155, 162, 163, 182].

The appearance of inversions is topologically excluded for the pair I (model 1, $l = 2$, $A \simeq B$) and for the pair VI (model 2, $l = 2$, $A \simeq B$), whilst all other pairs may have inversions within their TEMO patterns. This demand is satisfied with astonishing fidelity in view of the competition between topological and physical features. The characteristic interlacing of the MO's within various intervals is also exhibited; see for example the interval from 8.4 to 10.3 eV in the case of the pair III which contains 5 MO levels of S and T each. All these results may be rated as solid evidence for the physical relevance of TEMO.

The PE spectra of the pair II show a great number of inversions. From the characteristic polynomials of these topomers two inversions are deduced which are located below the HOMO and above the deepest π-MO, respectively. Since the energies of the levels at 9.01 and 9.06 eV are within the range of instrumental error, the two inversions near 9 eV are possibly due to experimental error. Thus, from the 6 inversions exhibited by the PE spectra of the topomers II there are at least two physically induced inversions.

Physically induced inversions have been investigated by means of variational perturbation theory at the level of non-empirical HF SCF MO calculations on much smaller molecules [148–151] than those shown in Fig. 13.1. The results obtained may be summarized as follows: (*i*) The variation of the potential at the various nuclei has effected the inversion only in the case of pyridazine (S) and pyrazine (T) [149]. (*ii*) All the other inversions found in the calculated MO spectra are due to interactions between non-neighbouring centers. It seems to be noteworthy that the use of an improved basis set reduces the number of inversions [151]. Hence, one could speculate that some of the inversions found by ab initio calculations might even be artefacts of the basis set used.

The results obtained quantum chemically support those derived from the PE spectra of pairs of topomers. Both taken together are a very solid evidence for the physical relevance of TEMO.

13.4 TEMO in σ-Electron Systems

As explained in Chap. 5 the hydrogen suppressed molecular graph used in HMO calculations is correctly understood as the graph which depicts the basis set of the p_π-AO's and their interactions [154]. The HMO method is characterized by defining an effective one electron operator upon this graph [169]. In order to construct a similar method for σ-electron systems, one could choose completely localized apolar σ- and σ^*-MO's for the basis. If it is assumed that two σ-MO's (σ^*-MO's) interact if and only if the corresponding two bonds have an atom in common and that all these interactions are equally valued, then the line graph of the skeleton graph of the molecule (see Chap. 3 and Sect. 4.4) is its σ-basis graph and an isomorphic line graph is the σ^*-basis graph [153]. Note that the vertices of these line graphs depict the σ- and σ^*-MO's, respectively. Their edges correspond to pairwise non-zero interactions of the MO's. If a line graph is derived from a graph which possesses vertices of degree $g \geq 3$, then it contains three-membered cycles; hence, the line graphs of interest will be non-bipartite.

Using hybrid orbitals (HO's) for the formation of the σ- and the σ^*-MO's one can show that an effective one electron operator $\mathcal{H}$ produces the following integrals

$$\alpha_j = \langle \sigma_j | \mathcal{H} | \sigma_j \rangle = \varepsilon_0 + \beta'$$
$$\alpha_j^* = \langle \sigma_j^* | \mathcal{H} | \sigma_j^* \rangle = \varepsilon_0 - \beta' \tag{13}$$

$$\beta_{jk} = \langle \sigma_j | \mathcal{H} | \sigma_k \rangle = \begin{cases} \beta & \text{if } j \text{ and } k \text{ are adjacent} \\ 0 & \text{otherwise} \end{cases} \tag{14}$$

$$\beta_{jk}^{**} = \langle \sigma_j^* | \mathcal{H} | \sigma_k^* \rangle = \begin{cases} \beta & \text{if } j \text{ and } k \text{ are adjacent} \\ 0 & \text{otherwise} \end{cases} \tag{15}$$

$$\beta_{jk}^* = \langle \sigma_j^* | \mathcal{H} | \sigma_k \rangle = 0 . \tag{16}$$

Proof. Let a, b, c, ... label subsequent trivalent centers of an unsaturated system; let the HO's located at these centers be a_i, b_i, c_i, ..., $i = 1, 2, 3$, and assume their mutual spatial orientation as follows:

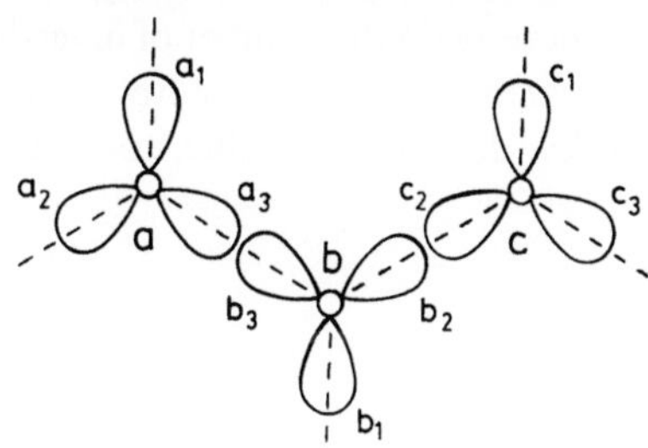

Then the non-polar σ- and σ^*-MO's for the bonds, say, j $(a -- b)$ and k $(b - c)$ are given by the following expressions

$$\sigma_j = N(a_3 + b_3) , \qquad \sigma_j^* = N^*(a_3 - b_3) \tag{17}$$

$$\sigma_k = N(b_2 + c_2) , \qquad \sigma_k^* = N^*(b_2 - c_2) \tag{18}$$

etc., where N and N^* denote the respective normalization constants.

Inserting these expressions into the integrals (13)–(16) one obtains

$$\varepsilon_0 = \langle a_i | \mathcal{H} | a_i \rangle = \langle b_i | \mathcal{H} | b_i \rangle = ... \tag{19}$$

where ε_0 is the energy of an electron in its so-called valence state [51],

$$\beta' = \langle a_3 | \mathcal{H} | b_3 \rangle = \langle b_2 | \mathcal{H} | c_2 \rangle = ... \tag{20}$$

where β' is the resonance integral of the two HO's which form the σ- or the σ^*-MO, and

$$\beta = \langle a_3 | \mathcal{H} | b_2 \rangle = \langle b_2 | \mathcal{H} | c_3 \rangle = ... \tag{21}$$

where β is the resonance integral of an HO of one center with any HO of an adjacent center, except that one with which it forms a σ- or a σ^*-MO.

All other integrals are zero either because of the orthogonality of the HO's of the same center or since all interactions between non-adjacent centers are neglected. $\square$

From (20) and (21) is seen that β' and β are "bonding" quantities, and that $|\beta'| > |\beta|$. In the case of saturated σ-electron systems the same formalism can be applied. The adopted energy equivalents of the parameters ε_0, β' and β will be slightly changed but not their formal meanings.

The various bonds as CH, CO, CC etc. are not differentiated; even lone pairs are treated as σ-MO's.

By means of this purely topological method similar results for σ-electron systems are obtained as described in the preceding paragraph for π-electron systems. In particular, TEMO without inversions results for the σ-systems of those topomers which are in accordance with either of the models 1 or 2 if $l = 2$ and $A \simeq B$.

Besides HMO-type assumptions, the topological σ-method requires some additional approximations. Thus the topological σ-method is not so close to physical reality as the corresponding HÜCKEL method for π-electrons and more violations of the TEMO rule should be expected in the case of σ-electron systems. This agrees with the results of quantum chemical calculations, but, here again the number of inversions is surprisingly small. Since the PE spectra of saturated compounds are not so nicely resolved as those of π-electron systems, only a few data are available, too few for evident conclusions.

13.5 TEMO and Symmetry

As already mentioned in Sect. 13.1, the isomorphism $A \simeq B$ induces some symmetry in the isomers S and T. In this very case one could think that perhaps the interlacing of their eigenvalues is induced by symmetry.

Let us consider for instance a pair of topomers constructed according to the model 1, $l = 2$, $A \simeq B$, as shown schematically in Fig. 13.2a. Due to the isomorphism $A \simeq B$, the MO's of A and B are pairwise equivalent, i.e. they have the same energies and the same systems of linear combinations of AO's (LCAO).

Applying the perturbation molecular orbital (PMO) theory [17, 35] to the i-th MO of A and B which have energy ε_i and LCAO MO coefficients $c_{ik} = c_{ip}$ and $c_{il} = c_{iq}$, one obtains the first-order perturbation energies $(c_{ik}^2 + c_{il}^2)\,\beta$ and $2c_{ik}c_{il}\beta$ for the topomers S and T, respectively. Evidently, $c_{ik}^2 + c_{il}^2 \geq 2c_{ik}c_{il}$ and thus the splitting of the levels ε_i in S and T meets one requirement of TEMO, namely the interlacing of two eigenvalues of T between two ones of S:

$$E_{2i-1}(S) = \varepsilon_i + (c_{ik}^2 + c_{il}^2)\,\beta$$

$$E_{2i-1}(T) = \varepsilon_i + 2c_{ik}c_{il}\beta$$

$$E_{2i}(T) = \varepsilon_i - 2c_{ik}c_{il}\beta \tag{22}$$

$$E_{2i}(S) = \varepsilon_i - (c_{ik}^2 + c_{il}^2)\,\beta$$

Table 13.3. PMO- and HMO-data for the pair VIII of topomers

i	ε_i	c_{ik} / c_{il}	$E_{2i-1}(S)$ / $E_{2i}(S)$	$E_{2i-1}(T)$ / $E_{2i}(T)$	$\lambda_j(S)$	$\lambda_j(T)$
			2.5800945		2.62946	
		0.1582147		2.5750834		2.61956
1	2.5474193					
		0.0874258		2.5197552		2.53281
			2.5147441		2.52791	
			2.1904652		2.26638	
		0.3093214		2.1762607		2.25874
2	2.0586327					
		0.1901388		1.9410047		2.00000
			1.9268002		1.97632	
			1.8275361		1.84199	
		0.0221996		1.8221690		1.83045
3	1.8179306					
		0.0954605		1.8136922		1.81662
			1.8083251		1.81194	
			1.6469045		1.64820	
		0.3360196		1.6441416		1.64770
4	1.4536477					
		0.2834566		1.2631538		1.44509
			ζ 1.2603909		1.43984	
			1.3456930		1.32286	
		0.2309918		1.3290783		1.31823
5	1.2819126					
		0.1020939		1.2347469		1.25894
			1.2181322		1.22857	
			1.0000000		1.00000	
		0.0000000		1.0000000		1.00000
6	1.0000000					
		0.0000000		1.0000000		1.00000
			ζ 1.0000000		1.00000	
			1.0364026		1.00000	
		0.2999756		1.0216656		1.00000
7	0.9145266					
		0.1785795		0.2073876		1.00000
			ζ 0.7926506		1.00000	
			0.8321162		0.76319	
		0.2776493		0.8221182		0.72753
8	0.6124155					
		0.3776395		0.4027128		0.50254
			ζ 0.3927148		0.50172	
			0.4545453		0.30664	
		0.3015113		0.3636381		0.25001
9	0.0000000					
		0.6030226		-0.3636381		-0.25001
			ζ -0.4545453		-0.30664	
etc.	etc.	etc.	etc.	etc.	etc.	etc.

where $E_j(S)$ and $E_j(T)$ denote the energy of the j-th MO of S and T, respectively, as calculated within the first-order PMO approximation.

The other demand of TEMO, namely the interlacing of the eigenvalues of S between the ones of T is not guaranteed by means of PMO. This is illustrated by Table 13.3 which collects the following data: ε_i, c_{ik} and c_{il} of the subunit A taken from [63], $E_j(S)$ and $E_j(T)$, and the HMO eigenvalues [138] $\lambda_j(S)$ and $\lambda_j(T)$, $j = 2i - 1, 2i$, of the topomers VIII S and VIII T

In Table 13.3 the disorders of the first-order PMO energies are marked by arrows. If these energies are ordered with respect to their magnitudes, readily done by the reader, the pattern obtained has four inversion points. In addition to this, one must recognize that in the case of non-isomorphic subunits, $A \neq B$, no conclusion analogous to (22) can be drawn from PMO theory. On the other hand, the TEMO rule is satisfied also in these cases, as shown by Tables 13.1 and 13.2.

All these facts give strong evidence that the interlacing is not induced by symmetry, but is a real consequence of molecular topology.

Appendices

In the first four appendices we shall briefly remind the reader to some of the most important definitions and theorems of matrix theory, theory of determinants, spectral theory of matrices and theory of polynomials. The facts outlined here should suffice for the reading of the present book. For more details and explanations the reader should, of course, consult an appropriate textbook.

In Appendix 5 we have collected the character tables of a number of point groups as well as of the first six symmetric groups. Appendix 6 provides a list of symbols used in the present book with a brief explanation and an indication where a more complete definition can be found.

Appendix 1

Matrices

The table M

$$
M = \begin{pmatrix}
m_{11} & m_{12} & \cdots & m_{1q} \\
m_{21} & m_{22} & \cdots & m_{2q} \\
\cdot & \cdot & \cdots & \cdot \\
m_{p1} & m_{p2} & \cdots & m_{pq}
\end{pmatrix} \tag{1}
$$

composed of $p \cdot q$ numbers, arranged in p rows and q columns is called a *matrix*. Such a matrix is of *dimension* $p \times q$ and the numbers which form the matrix are its *elements*. The matrix element m_{ij} is said to be in the i-th row and in the j-th column. If $p = 1$, then the matrix

$$
(m_{11} m_{12} \ldots m_{1q}) \tag{2}
$$

is called *row-vector* of dimension q. If $q = 1$, then the matrix

$$
\begin{pmatrix}
m_{11} \\
m_{21} \\
\vdots \\
m_{p1}
\end{pmatrix} \tag{3}
$$

is called a *column-vector* of dimension p. If $p = q$, then we have a *square* matrix of order p.

Let N be another matrix of dimension $r \times s$,

$$
N = \begin{pmatrix}
n_{11} & n_{12} & \cdots & n_{1s} \\
n_{21} & n_{22} & \cdots & n_{2s} \\
\cdot & \cdot & \cdots & \cdot \\
n_{r1} & n_{r2} & \cdots & n_{rs}
\end{pmatrix}. \tag{4}
$$

Then the sum of the matrices M and N exists if $p = r$ and $q = s$ and is given by

$$
M + N = \begin{pmatrix}
m_{11} + n_{11} & m_{12} + n_{12} & \cdots & m_{1q} + n_{1q} \\
m_{21} + n_{21} & m_{22} + n_{22} & \cdots & m_{2q} + n_{2q} \\
\cdot & \cdot & \cdots & \cdot \\
m_{p1} + n_{p1} & m_{p2} + n_{p2} & \cdots & m_{pq} + n_{pq}
\end{pmatrix}. \tag{5}
$$

The difference $M - N$ is defined analogously. If two matrices are not of the same dimension, their sum and difference are not defined.

If c is a number, then the product of c and M is determined as

$$cM = \begin{pmatrix} cm_{11} & cm_{12} & \cdots & cm_{1q} \\ cm_{21} & cm_{22} & \cdots & cm_{2q} \\ \cdot & \cdot & \cdots & \cdot \\ cm_{p1} & cm_{p2} & \cdots & cm_{pq} \end{pmatrix}. \tag{6}$$

A matrix all of whose elements are equal to zero is the *zero matrix* and will be denoted by O. It is clear that $M + O = M$ and $cO = O$.

The product T of two matrices M and N of dimension $p \times q$ and $r \times s$, respectively, exists only if $q = r$ and is defined as a matrix of dimension $p \times s$

$$T = \begin{pmatrix} t_{11} & t_{12} & \cdots & t_{1s} \\ t_{21} & t_{22} & \cdots & t_{2s} \\ \cdot & \cdot & \cdots & \cdot \\ t_{p1} & t_{p2} & \cdots & t_{ps} \end{pmatrix} \tag{7}$$

such that

$$t_{ij} = \sum_{k=1}^{q} m_{ik} n_{kj} . \tag{8}$$

If the elements of the matrix T are defined via (8), then we shall write $T = MN$. The product $T' = NM$ exists only if $p = s$ and its matrix elements obey

$$t'_{ij} = \sum_{k=1}^{p} n_{ik} m_{kj} . \tag{9}$$

Both the products MN and NM exist only if the conditions $q = r$ and $p = s$ are simultaneously fulfilled. In particular, both MN and NM exist if M and N are square matrices of the same order. In the general case MN differs from NM, as shown by the following example.

Let

$$M = \begin{pmatrix} 1 & 2 \\ 3 & 4 \end{pmatrix} \qquad N = \begin{pmatrix} 5 & 6 \\ 7 & 8 \end{pmatrix} \tag{10}$$

and $T = MN$. Then

$$t_{11} = 1 \cdot 5 + 2 \cdot 7 = 19$$

$$t_{12} = 1 \cdot 6 + 2 \cdot 8 = 22$$

$$t_{21} = 3 \cdot 5 + 4 \cdot 7 = 43 \tag{11}$$

$$t_{22} = 3 \cdot 6 + 4 \cdot 8 = 50$$

and hence

$$MN = \begin{pmatrix} 19 & 22 \\ 43 & 50 \end{pmatrix}. \tag{12}$$

If $T' = NM$, then

$$\begin{aligned}
t'_{11} &= 5 \cdot 1 + 6 \cdot 3 = 23 \\
t'_{12} &= 5 \cdot 2 + 6 \cdot 4 = 34 \\
t'_{21} &= 7 \cdot 1 + 8 \cdot 3 = 31 \\
t'_{22} &= 7 \cdot 2 + 8 \cdot 4 = 46
\end{aligned} \tag{13}$$

and

$$NM = \begin{pmatrix} 23 & 34 \\ 31 & 46 \end{pmatrix}. \tag{14}$$

Consequently $MN \neq NM$.

If $MN = NM$ then the matrices M and N are said to commute.

In the following we shall be concerned mainly with square matrices. If not otherwise stated the matrices considered are square of order n.

Let M be a square matrix of order n. Then the matrix elements m_{11}, m_{22}, ... , m_{nn} form the *diagonal of* M. The numbers m_{ij}, $i \neq j$ are the off-diagonal elements of M.

The *trace* of M is the sum of its diagonal elements:

$$\mathrm{Tr}\ M = \sum_{i=1}^{n} m_{ii}. \tag{15}$$

If all the off-diagonal elements of M are equal to zero, then M is said to be a *diagonal matrix* and we shall write

$$M = \mathrm{diag}\,(m_{11}, m_{22}, ... , m_{nn}). \tag{16}$$

The diagonal matrix whose diagonal elements are equal to unity is called the *unit matrix* and will be denoted by I_n or (where there is no danger of misunderstanding) simply by I.

The unit matrix has the obvious property

$$MI = IM = M \tag{17}$$

for all square matrices M of the same order as that of I.

If $MN = I$ then N is the inverse matrix of M and is usually denoted by M^{-1}. A necessary and sufficient condition for the existence of M^{-1} is that the determinant of M is non-zero. The elements of M^{-1} are given by Eq. (13) of Appendix 2.

If M and N are matrices such that for all i, j,

$$n_{ij} = m_{ji} \tag{18}$$

then N is the *transpose* of M and will be denoted by $M^\dagger$. If M is a square matrix and $m_{ij} = m_{ji}$, then M is a *symmetric matrix*. Hence a matrix M is symmetric if $M = M^\dagger$.

If m_{ij} is the complex conjugate of m_{ji}, then M is said to be a *Hermitean matrix*. A symmetric matrix whose elements are real numbers is Hermitean.

If a matrix U has the properties

$$UU^\dagger = U^\dagger U = I \tag{19}$$

then it is called *unitary*. Hence U is a unitary matrix if $U^\dagger = U^{-1}$.

The reader can easily verify that

$$\begin{pmatrix} \cos \varphi & -\sin \varphi \\ \sin \varphi & \cos \varphi \end{pmatrix} \tag{20}$$

is a unitary matrix. This example is given in order to illustrate the following important application of matrices.

Consider a point (x, y) in a two-dimensional coordinate system. If we rotate this point by an angle φ around the origin of the coordinate system then the new coordinates (x', y') satisfy the equation

$$\begin{pmatrix} x' \\ y' \end{pmatrix} = \begin{pmatrix} \cos \varphi & -\sin \varphi \\ \sin \varphi & \cos \varphi \end{pmatrix} \begin{pmatrix} x \\ y \end{pmatrix}. \tag{21}$$

Let M, N and U be square matrices of the same order. Let in addition the determinant of U be non-zero. Then U^{-1} exists.

If $N = U^{-1}MU$, then the matrices N and M are said to be *similar* and the mapping of M into N is a *similarity transformation*. An important property of similar matrices is that their traces are equal. Hence

$$\sum_i (M)_{ii} = \sum_i (U^{-1}MU)_{ii} \tag{22}$$

for all matrices U which have an inverse.

In particular, if U is a unitary matrix, then we speak about *unitary transformations*. Unitary matrices and unitary transformations play an important role in the spectral theory of matrices (see Appendix 3).

A matrix

$$M = \begin{pmatrix}
m_{11} & m_{12} & \cdots & m_{1q} & m_{1,q+1} & m_{1,q+2} & \cdots & m_{1,q+s} \\
m_{21} & m_{22} & \cdots & m_{2q} & m_{2,q+1} & m_{2,q+2} & \cdots & m_{2,q+s} \\
\cdot & \cdot & \cdots & \cdot & \cdot & \cdot & \cdots & \cdot \\
m_{p1} & m_{p2} & \cdots & m_{pq} & m_{p,q+1} & m_{p,q+2} & \cdots & m_{p,q+s} \\
m_{p+1,1} & m_{p+1,2} & \cdots & m_{p+1,q} & m_{p+1,q+1} & m_{p+1,q+2} & \cdots & m_{p+1,q+s} \\
m_{p+2,1} & m_{p+2,2} & \cdots & m_{p+2,q} & m_{p+2,q+1} & m_{p+2,q+2} & \cdots & m_{p+2,q+s} \\
\cdot & \cdot & \cdots & \cdot & \cdot & \cdot & \cdots & \cdot \\
m_{p+r,1} & m_{p+r,2} & \cdots & m_{p+r,q} & m_{p+r,q+1} & m_{p+r,q+2} & \cdots & m_{p+r,q+s}
\end{pmatrix} \tag{23}$$

can be viewed as composed of the blocks

$$M_1 = \begin{pmatrix} m_{11} & m_{12} & \cdots & m_{1q} \\ m_{21} & m_{22} & \cdots & m_{2q} \\ \cdot & \cdot & \cdots & \cdot \\ m_{p1} & m_{p2} & \cdots & m_{pq} \end{pmatrix} \tag{24}$$

$$M_2 = \begin{pmatrix} m_{1,q+1} & m_{1,q+2} & \cdots & m_{1,q+s} \\ m_{2,q+1} & m_{2,q+2} & \cdots & m_{2,q+s} \\ \cdot & \cdot & \cdots & \cdot \\ m_{p,q+1} & m_{p,q+2} & \cdots & m_{p,q+s} \end{pmatrix} \tag{25}$$

$$M_3 = \begin{pmatrix} m_{p+1,1} & m_{p+1,2} & \cdots & m_{p+1,q} \\ m_{p+2,1} & m_{p+2,2} & \cdots & m_{p+2,q} \\ \cdot & \cdot & \cdots & \cdot \\ m_{p+r,1} & m_{p+r,2} & \cdots & m_{p+r,q} \end{pmatrix} \tag{26}$$

$$M_4 = \begin{pmatrix} m_{p+1,q+1} & m_{p+1,q+2} & \cdots & m_{p+1,q+s} \\ m_{p+2,q+1} & m_{p+2,q+2} & \cdots & m_{p+2,q+s} \\ \cdot & \cdot & \cdots & \cdot \\ m_{p+r,q+1} & m_{p+r,q+2} & \cdots & m_{p+r,q+s} \end{pmatrix} \tag{27}$$

that is

$$M = \begin{pmatrix} M_1 & M_2 \\ M_3 & M_4 \end{pmatrix}. \tag{28}$$

Of particular importance are the matrices having the so-called *block diagonal* form:

$$M = \begin{pmatrix} M_1 & 0 \\ 0 & M_2 \end{pmatrix} \tag{29}$$

where M_1 and M_2 are square matrices. A frequently used property of such matrices is that the determinant of M is equal to the product of the determinants of M_1 and M_2.

Appendix 2

Determinants

Let $\mathscr{I}_n$ be the set of the first n positive integers:

$$\mathscr{I}_n = \{1, 2, \ldots, n\} \tag{1}$$

Every mapping P of $\mathscr{I}_n$ onto itself is called[1] a *permutation*. We shall present the mapping P as

$$P: \begin{pmatrix} 1 & 2 & \ldots & n \\ k_1 & k_2 & \ldots & k_n \end{pmatrix} \tag{2}$$

which means that P maps 1 into k_1, 2 into $k_2, \ldots n$ into k_n and furthermore $\{k_1, k_2, \ldots, k_n\} = \mathscr{I}_n$.

It is well known that there are $n!$ distinct permutations of n elements. A permutation of the form

$$\begin{pmatrix} 1 & 2 & \ldots & i-1 & i & i+1 & \ldots & j-1 & j & j+1 & \ldots & n \\ 1 & 2 & \ldots & i-1 & j & i+1 & \ldots & j-1 & i & j+1 & \ldots & n \end{pmatrix} \tag{3}$$

which interchanges a single pair of numbers, i and j, is called a *transposition*. Any permutation P can be obtained by making a certain number of successive transpositions. Let the number of transpositions required for the creation of the permutation P be denoted by $r(P)$. Then the term $(-1)^{r(P)}$ is a unique characteristic of the permutation P and is called its *parity*.

Let M be a square matrix of order n. The *determinant* of M is defined as

$$\det M = \sum_P (-1)^{r(P)} m_{1k_1} m_{2k_2} \cdots m_{nk_n} \tag{4}$$

where the summation goes over all $n!$ permutations of the numbers $1, 2, \ldots, n$. In particular, for $n = 1$

$$\det (m_{11}) = m_{11} \tag{5}$$

---

[1] A more detailed account of the theory of permutations is given in Chap. 9.

for $n = 2$,

$$\det \begin{pmatrix} m_{11} & m_{12} \\ m_{21} & m_{22} \end{pmatrix} = m_{11}m_{22} - m_{12}m_{21} \tag{6}$$

for $n = 3$,

$$\det \begin{pmatrix} m_{11} & m_{12} & m_{13} \\ m_{21} & m_{22} & m_{23} \\ m_{31} & m_{32} & m_{33} \end{pmatrix} = m_{11}m_{22}m_{33} - m_{11}m_{23}m_{32}$$

$$+ \, m_{12}m_{23}m_{31} - m_{12}m_{21}m_{33} + m_{13}m_{21}m_{32} - m_{13}m_{22}m_{31} \,. \tag{7}$$

Let M_{ij} be the matrix obtained by deleting from M the i-th row and the j-th column. Then the *cofactor* of the element m_{ij} of M is

$$\mu_{ij} = (-1)^{i+j} \det M_{ij} \,. \tag{8}$$

The following results concerning the cofactors are worth mentioning.
1° If M is symmetric, then $\mu_{ij} = \mu_{ji}$.
2° Since it is easier to calculate the cofactors than the determinant itself, the following expansion formulas

$$\det M = \sum_{i=1}^{n} m_{ij}\mu_{ij} = \sum_{i=1}^{n} m_{ji}\mu_{ji} \tag{9}$$

are often used in practice. They hold for all $j \in \mathscr{I}_n$. If $k \neq j$, then

$$\sum_{i=1}^{n} m_{ij}\mu_{ik} = \sum_{i=1}^{n} m_{ji}\mu_{ki} = 0 \,. \tag{10}$$

3° From (9) it is evident that

$$\frac{\partial \det M}{\partial m_{ij}} = \mu_{ij} \,. \tag{11}$$

Furthermore, if M is a matrix whose diagonal elements are all equal to x, then

$$\frac{\partial \det M}{\partial x} = \sum_{i=1}^{n} \mu_{ii} \,. \tag{12}$$

4° The inverse of M exists if and only if $\det M \neq 0$ and then

$$(M^{-1})_{ij} = (\det M)^{-1} \mu_{ji} \,. \tag{13}$$

5° Denote by $M|12 \ldots q$ the matrix obtained by deletion from M the first q rows and the first q columns. Then

$$
\det \begin{pmatrix} \mu_{11} & \mu_{12} & \cdots & \mu_{1q} \\ \mu_{21} & \mu_{22} & \cdots & \mu_{2q} \\ \cdot & \cdot & \cdots & \cdot \\ \mu_{q1} & \mu_{q2} & \cdots & \mu_{qq} \end{pmatrix} = (\det M)^{q-1} \det (M|12\ldots q). \tag{14}
$$

For $q = 2$ we have the following special case of (14):

$$
\det (M|i) \det (M|j) - \det M \det (M|ij) = \mu_{ij}\mu_{ji} . \tag{15}
$$

Formulas (14) and (15) are known as the JACOBI *identity*. Note that $\det (M|i) = \mu_{ii}$.

Among many other properties of determinants we mention here the formula (16) because it will be needed later on:

$$
\det \begin{pmatrix} m_{11} + m'_{11} & m_{12} + m'_{12} & \cdots & m_{1n} + m'_{1n} \\ m_{21} & m_{22} & \cdots & m_{2n} \\ \cdot & \cdot & \cdots & \cdot \\ m_{n1} & m_{n2} & \cdots & m_{nn} \end{pmatrix}
$$

$$
= \det \begin{pmatrix} m_{11} & m_{12} & \cdots & m_{1n} \\ m_{21} & m_{22} & \cdots & m_{2n} \\ \cdot & \cdot & \cdots & \cdot \\ m_{n1} & m_{n2} & \cdots & m_{nn} \end{pmatrix} + \det \begin{pmatrix} m'_{11} & m'_{12} & \cdots & m'_{1n} \\ m_{21} & m_{22} & \cdots & m_{2n} \\ \cdot & \cdot & \cdots & \cdot \\ m_{n1} & m_{n2} & \cdots & m_{nn} \end{pmatrix}. \tag{16}
$$

Two determinants of order n can be added (in the sense of the above equation) only if $n - 1$ of their rows or $n - 1$ of their columns coincide.

Appendix 3

Eigenvalues and Eigenvectors

In the present appendix M will always denote a real and symmetric square matrix of order n. In other words the elements of M are real numbers and $m_{ij} = m_{ji}$.

According to the definition of matrix multiplication, if C is an n-dimensional column-vector, then MC is also an n-dimensional column-vector.

Two vectors C and C' are said to be *collinear* if there is a constant λ, such that $C' = \lambda C$. Those vectors C which have the property that C and MC are collinear are the *eigenvectors* of the matrix M. The corresponding constants λ are the *eigenvalues* of M. The equation

$$MC = \lambda C \tag{1}$$

is called the eigenvalue-eigenvector equation of the matrix M.

It can be shown that there exist n linearly independent vectors $C_1, C_2, \dots, C_n$ which satisfy Eq. (1). The corresponding eigenvalues (which need not be all distinct) will be labeled by $\lambda_1, \lambda_2, \dots, \lambda_n$.

If the (linearly independent) eigenvectors $C_{i_1}, C_{i_2}, \dots, C_{i_d}$ have the same eigenvalue (i.e. $\lambda = \lambda_{i_1} = \lambda_{i_2} = \dots = \lambda_{i_d}$), then they are d-times degenerate and the eigenvalue λ is said to have algebraic multiplicity d.

Two eigenvectors C_i and C_j of M which correspond to different eigenvalues (which are not degenerate, $\lambda_i \neq \lambda_j$) must be *orthogonal*:

$$C_i^\dagger C_j = 0 . \tag{2}$$

Degenerate eigenvectors need not be orthogonal. However, if C_i and C_j are degenerate eigenvectors of M, then their arbitrary linear combination (that is $pC_i + qC_j$, where p and q are arbitrary numbers) is also an eigenvector of M with the same eigenvalue. Using this property of degenerate eigenvectors, one can always construct their linear combinations which are mutually orthogonal.

In addition to this, the eigenvectors of M can be chosen so as to be *normalized*:

$$C_i^\dagger C_i = 1 \tag{3}$$

for all i.

If all the eigenvectors are normalized and mutually orthogonal, then they are said to be *orthonormal*. The orthonormality of the eigenvectors can be presented in the form

$$C_i^\dagger C_j = \delta_{ij} \tag{4}$$

where the symbol δ_{ij} is the so called "KRONECKER delta", defined as $\delta_{ij} = 0$ if $i \neq j$ and $\delta_{ii} = 1$.

For every matrix M with the above described properties a unitary matrix U can be found such that

$$U^{-1}MU = \mathrm{diag}\,(\lambda_1, \lambda_2, \ldots, \lambda_n)\,. \tag{5}$$

The diagonal elements of $U^{-1}MU$ are just the eigenvalues of M. All eigenvalues of M, with pertinent algebraic multiplicities, appear on the diagonal of $U^{-1}MU$. Furthermore,

$$U = (C_1, C_2, \ldots, C_n)\,. \tag{6}$$

The unitary matrix U satisfying (5), is said to *diagonalize* the matrix M. Two matrices are diagonalized by the same unitary matrix if and only if they commute. This means that matrices which commute have the same eigenvectors.

The *characteristic polynomial* of the matrix M is defined as

$$\varphi(M, x) = \det\,(xI - M)\,. \tag{7}$$

Bearing in mind the properties of the determinant, one immediately sees that $\varphi(M, x)$ is a polynomial of degree n (in the variable x). The eigenvalues of M are the zeros of $\varphi(M, x)$, that is

$$\varphi(M, \lambda_i) = 0 \tag{8}$$

for all $i = 1, 2, \ldots, n$. If λ is an eigenvalue of a d-times degenerate eigenvector, then λ is a d-fold zero of $\varphi(M, x)$.

The numbers $\lambda_1, \lambda_2, \ldots, \lambda_n$ (taking into account their algebraic multiplicities) form the *spectrum* of the matrix M.

If M is a real and symmetric (or more general: a Hermitean) matrix, then its spectrum is composed of real numbers.

Define $M^2 = MM$, $M^3 = M^2M$ etc. It can be shown that for all $k \geq 1$,

$$\mathrm{Tr}\,(M^k) = \sum_{i=1}^{n} \lambda_i^k\,. \tag{9}$$

The expression on the right-hand side of (9) is sometimes called the k-th moment of the matrix M.

Let us write the characteristic polynomial (7) as

$$x^n + a_1 x^{n-1} + a_2 x^{n-2} + \dots + a_n . \tag{10}$$

Then the CAYLEY-HAMILTON theorem claims that

$$M^n + a_1 M^{n-1} + a_2 M^{n-2} + \dots + a_n I = 0 . \tag{11}$$

Appendix 4

Polynomials

Let $a_0, a_1, \ldots, a_{n-1}, a_n$ be a sequence of $n+1$ numbers, $n > 0$. Then the expression

$$P(x) = a_0 x^n + a_1 x^{n-1} + \ldots + a_{n-1} x + a_n \tag{1}$$

is called a *polynomial* in the variable x. If $a_0 \neq 0$, then $P(x)$ is a polynomial of degree n.
The numbers $a_0, a_1, \ldots, a_n$ are the coefficients of $P(x)$.
Let $Q(x)$ be another polynomial

$$Q(x) = b_0 x^m + b_1 x^{m-1} + \ldots + b_{m-1} x + b_m . \tag{2}$$

The polynomials $P(x)$ and $Q(x)$ are equal if $n = m$ and $a_i = b_i$ for all i. Then, of course, the equality $P(x) = Q(x)$ holds for all x. Whenever there is no danger of misunderstanding, we shall denote the fact that two polynomials $P(x)$ and $Q(x)$ are equal simply by $P(x) = Q(x)$.

If for $x = \lambda$ the value of the polynomial $P(x)$ is equal to zero, then λ is a *zero of the polynomial* $P(x)$. Hence λ is a *root of the equation* $P(x) = 0$.

A fundamental theorem of algebra is that every polynomial $P_n(x)$ of degree $n > 0$ has at least one zero and, consequently, can be written in the form

$$P_n(x) = (x - \lambda_1) P_{n-1}(x) \tag{3}$$

where $P_{n-1}(x)$ is a polynomial of degree $n - 1$ (if $n > 1$) or a constant (if $n = 1$). In Eq. (3) the zero of $P_n(x)$ is denoted by λ_1. This can be either a real or a complex number.

A proper consequence of (3) is that a polynomial $P(x)$ of degree n can be written in the form

$$P(x) = a_0(x - \lambda_1)(x - \lambda_2) \ldots (x - \lambda_n) \tag{4}$$

where $\lambda_1, \lambda_2, \ldots, \lambda_n$ are its zeros. A certain zero of $P(x)$ can occur several times in the product on the right-hand side of (4). If $\lambda_{i_1} = \lambda_{i_2} = \ldots = \lambda_{i_d} = \lambda$, then we say that λ is a d-fold zero of $P(x)$ or that the *algebraic multiplicity* of λ is equal to d or that its *degeneracy* is equal to d.

If the algebraic multiplicity is taken into account, then every polynomial of degree n has exactly n zeros.

The coefficients and the zeros of a polynomial are related by the so-called VIETA *formulas*. Let $\lambda_1, \lambda_2, \dots, \lambda_n$ be the zeros of the polynomial $P(x)$ defined via Eq. (1). Then

$$-a_1/a_0 = \sum_{i=1}^{n} \lambda_i \tag{5}$$

$$+a_2/a_0 = \sum_{i<j} \lambda_i \lambda_j \tag{6}$$

$$-a_3/a_0 = \sum_{i<j<k} \lambda_i \lambda_j \lambda_k \tag{7}.$$

etc.

Define the quantity S_k (which is sometimes called the k-th moment) as

$$S_k = \sum_{i=1}^{n} \lambda_i^k . \tag{8}$$

Then the relations between S_k and the coefficients of the polynomial $P(x)$ are given by the NEWTON *identities*:

$$S_k a_0 + S_{k-1} a_1 + \dots + S_1 a_{k-1} + k a_k = 0 \tag{9}$$

if $k \leqq n$, whereas for $k > n$,

$$S_k a_0 + S_{k-1} a_1 + \dots + S_{k-n} a_n = 0 . \tag{10}$$

It would be useful to compare Eq. (9) from Appendix 3 with the present Eq. (8). Relations (9) and (10) hold for the moments of a matrix and the coefficients of its characteristic polynomial.

Polynomials whose zeros are real deserve a special attention because of their role in the present book. For instance, all the zeros of the characteristic polynomial of a symmetric real matrix (and therefore of the characteristic polynomial of a graph) are real. The same is true for the matching polynomial.

For such polynomials the DESCARTES *theorem* enables one to determine the number of positive zeros:

Let a polynomial $P(x)$ be given by Eq. (1) and let all its zeros be real numbers. Then the number n_+ of positive zeros of $P(x)$ is equal to the number of sign changes in the sequence $a_0, a_1, \dots, a_n$. Coefficients which are equal to zero do not contribute to sign changes.

If in (1) $a_k \neq 0$ and $a_{k+1} = a_{k+2} = \dots = a_n = 0$, then $\lambda = 0$ is an $(n - k)$-fold zero of $P(x)$. Hence $P(x)$ has $n_0 = n - k$ zeros which are equal to zero. Since the total number of zeros of $P(x)$ is equal to n, the number n_- of negative zeros is equal to $n - n_+ - n_0$. Consequently, from the knowledge of the coefficients $a_0, a_1, \dots, a_n$ one can easily deduce the number of positive, zero and negative zeros of $P(x)$.

Appendix 5

Characters of Irreducible Representations of Symmetry Groups

In this appendix we present the character tables of the symmetry groups which are most frequently used in organic chemistry. In the left column of these tables the irreducible representations of the group are indicated. The next column gives the respective characters for the different classes of symmetry elements. In the right column the transformation properties of some functions under the symmetry operations of the group are indicated. R_x, R_y and R_z denote the rotation of the molecule around the axes of the coordinate system (see Eq. (8.27)). The transformation properties (see Sect. 8.3) of the functions x, y, z, xy, yz and zx are always in harmony with a particular irreducible representation. In some groups, however, the transformation behaviour of the functions x^2, y^2 and z^2 cannot be properly associated with any irreducible representation. In such cases, linear combinations like $x^2 + y^2$, $x^2 - y^2$ etc. are considered, which transform in accordance with a particular representation.

For further notation used in the following tables the reader should consult Chap. 8. The character tables of some further symmetry groups can be found in [58].

We give here also the character tables of the symmetric groups S_n, $n \leq 6$ [41]. The group elements are denoted by their cycle structures. Below each cycle structure the number of elements of the corresponding class is given. For details see Chap. 9.

C_1	E			
A	1	x, y, z	R_x, R_y, R_z	$x^2, y^2, z^2, xy, yz, zx$

C_2	E	C_2			
A	1	1	z	R_z	x^2, y^2, z^2, xy
B	1	-1	x, y	R_x, R_y	yz, zx

C_3	E	C_3	C_3^2	$\omega = \exp(2\pi i/3)$		
A	1	1	1	z	R_z	$x^2 + y^2, z^2$
E $\left\{\begin{array}{c} \\ \end{array}\right.$	$\begin{array}{c}1\\1\end{array}$	$\begin{array}{c}\omega\\\omega^*\end{array}$	$\begin{array}{c}\omega^2\\\omega^{*2}\end{array}$ $\Big\}$	x, y	R_x, R_y $\Big\}$	$x^2 - y^2, xy,$ yz, zx $\Big\}$

C_4	E	C_4	C_2	C_4^3			
A	1	1	1	1	z	R_z	$x^2 + y^2, z^2$
B	1	-1	1	-1			$x^2 - y^2, xy$
E $\left\{\begin{array}{l}\\ \\\end{array}\right.$	$\begin{array}{l}1\\1\end{array}$	$\begin{array}{l}i\\-i\end{array}$	$\begin{array}{l}-1\\-1\end{array}$	$\begin{array}{l}-i\\i\end{array}$	$\left.\begin{array}{l}\\ \\\end{array}\right\}$ x, y	$\left.\begin{array}{l}\\ \\\end{array}\right\}$ R_x, R_y	$\left.\begin{array}{l}\\ \\\end{array}\right\}$ yz, zx

C_5	E	C_5	C_5^2	C_5^3	C_5^4	$\omega = \exp(2\pi i/5)$		
A	1	1	1	1	1	z	R_z	$x^2 + y^2, z^2$
E_1 $\left\{\begin{array}{l}\\ \\\end{array}\right.$	$\begin{array}{l}1\\1\end{array}$	$\begin{array}{l}\omega\\\omega^*\end{array}$	$\begin{array}{l}\omega^2\\\omega^{*2}\end{array}$	$\begin{array}{l}\omega^3\\\omega^{*3}\end{array}$	$\begin{array}{l}\omega^4\\\omega^{*4}\end{array}$	$\left.\begin{array}{l}\\ \\\end{array}\right\}$ x, y	$\left.\begin{array}{l}\\ \\\end{array}\right\}$ R_x, R_y	$\left.\begin{array}{l}\\ \\\end{array}\right\}$ yz, zx
E_2 $\left\{\begin{array}{l}\\ \\\end{array}\right.$	$\begin{array}{l}1\\1\end{array}$	$\begin{array}{l}\omega^2\\\omega^{*2}\end{array}$	$\begin{array}{l}\omega^4\\\omega^{*4}\end{array}$	$\begin{array}{l}\omega\\\omega^*\end{array}$	$\begin{array}{l}\omega^3\\\omega^{*3}\end{array}$			$\left.\begin{array}{l}\\ \\\end{array}\right\}$ $x^2 - y^2, xy$

C_6	E	C_6	C_3	C_2 C_3^2		C_6^5	$\omega = \exp(2\pi i/6)$		
A	1	1	1	1	1	1	z	R_z	$x^2 + y^2, z^2$
B	1	-1	1	-1	1	-1			
E_1 $\left\{\begin{array}{l}\\ \\\end{array}\right.$	$\begin{array}{l}1\\1\end{array}$	$\begin{array}{l}\omega\\\omega^*\end{array}$	$\begin{array}{l}\omega^2\\\omega^{*2}\end{array}$	$\begin{array}{l}-1\\-1\end{array}$	$\begin{array}{l}-\omega\\-\omega^*\end{array}$	$\begin{array}{l}-\omega^2\\-\omega^{*2}\end{array}$	$\left.\begin{array}{l}\\ \\\end{array}\right\}$ x, y	$\left.\begin{array}{l}\\ \\\end{array}\right\}$ R_x, R_y	$\left.\begin{array}{l}\\ \\\end{array}\right\}$ yz, zx
E_2 $\left\{\begin{array}{l}\\ \\\end{array}\right.$	$\begin{array}{l}1\\1\end{array}$	$\begin{array}{l}\omega^2\\\omega^{*2}\end{array}$	$\begin{array}{l}\omega^4\\\omega^{*4}\end{array}$	$\begin{array}{l}1\\1\end{array}$	$\begin{array}{l}\omega^2\\\omega^{*2}\end{array}$	$\begin{array}{l}\omega^4\\\omega^{*4}\end{array}$			$\left.\begin{array}{l}\\ \\\end{array}\right\}$ $x^2 - y^2, xy$

C_{2v}	E	C_2	σ_v zx	σ_d yz			
A_1	1	1	1	1	z		x^2, y^2, z^2
A_2	1	1	-1	-1		R_z	xy
B_1	1	-1	1	-1	x	R_y	zx
B_2	1	-1	-1	1	y	R_x	yz

C_{3v}	E	$2C_3$	$3\sigma_v$			
A_1	1	1	1	z		$x^2 + y^2, z^2$
A_2	1	1	-1		R_z	
E	2	-1	0	x, y	R_x, R_y	$\left\{\begin{array}{l}x^2 - y^2, xy\\yz, zx\end{array}\right.$

C_{4v}	E	$2C_4$	C_2	$2\sigma_v$	$2\sigma_d$			
A_1	1	1	1	1	1	z		$x^2+y^2,\ z^2$
A_2	1	1	1	-1	-1		R_z	
B_1	1	-1	1	1	-1			x^2-y^2
B_2	1	-1	1	-1	1			xy
E	2	0	-2	0	0	x,y	R_x, R_y	$yz,\ zx$

C_{5v}	E	$2C_5$	$2C_5^2$	$5\sigma_v$	$\varphi=2\pi/5$		
A_1	1	1	1	1	z		$x^2+y^2,\ z^2$
A_2	1	1	1	-1		R_z	
E_1	2	a	b	0	x,y	R_x, R_y	$yz,\ zx$
E_2	2	b	a	0			$x^2-y^2,\ xy$

$a = 2\cos\varphi = 2\cos 4\varphi = (\sqrt{5}-1)/2 = 0.618034,$

$b = 2\cos 2\varphi = -(\sqrt{5}+1)/2 = -1.618034.$

C_{6v}	E	$2C_6$	$2C_3$	C_2	$3\sigma_v$	$3\sigma_d$			
A_1	1	1	1	1	1	1	z		$x^2+y^2,\ z^2$
A_2	1	1	1	1	-1	-1		R_z	
B_1	1	-1	1	-1	1	-1			
B_2	1	-1	1	-1	-1	1			
E_1	2	1	-1	-2	0	0	x,y	R_x, R_y	$yz,\ zx$
E_2	2	-1	-1	2	0	0			$x^2-y^2,\ xy$

C_{1h}	E	σ_h	$C_{1h}=C_s$		
		xy			
A'	1	1	x,y	R_z	x^2, y^2, z^2, xy
A''	1	-1	z	R_x, R_y	$yz,\ zx$

C_{2h}	E	C_2	i	σ_h			
A_g	1	1	1	1		R_z	x^2, y^2, z^2, xy
A_u	1	1	-1	-1	z		
B_g	1	-1	1	-1		R_x, R_y	$yz,\ zx$
B_u	1	-1	-1	1	x,y		

C_{3h}	E	C_3	C_3^2	σ_h	S_3	S_3^2			
A'	1	1	1	1	1	1		R_z	$x^2 + y^2,\ z^2$
A''	1	1	1	-1	-1	-1	z		
$E'\ \Big\{$	1	ω	ω^2	1	ω	ω^2	$\Big\}\,x, y$		$\Big\}\,x^2 - y^2,\ xy$
	1	ω^*	ω^{*2}	1	ω^*	ω^{*2}			
$E''\ \Big\{$	1	ω	ω^2	-1	$-\omega$	$-\omega^2$		$\Big\}\,R_x, R_y$	$\Big\}\,yz, zx$
	1	ω^*	ω^{*2}	-1	$-\omega^*$	$-\omega^{*2}$			

C_{4h}	E	C_4	C_2	C_4^3	i	S_4^3	σ_h	S_4			
A_g	1	1	1	1	1	1	1	1		R_z	$x^2 + y^2,\ z^2$
A_u	1	1	1	1	-1	-1	-1	-1	z		
B_g	1	-1	1	-1	1	-1	1	-1			$x^2 - y^2,\ xy$
B_u	1	-1	1	-1	-1	1	-1	1			
$E_g\ \Big\{$	1	i	-1	$-i$	1	i	-1	$-i$		$\Big\}\,R_x, R_y$	$\Big\}\,yz, zx$
	1	$-i$	-1	i	1	$-i$	-1	i			
$E_u\ \Big\{$	1	i	-1	$-i$	-1	$-i$	1	i	$\Big\}\,x, y$		
	1	$-i$	-1	i	-1	i	1	$-i$			

C_{5h}	E	C_5	C_5^2	C_5^3	C_5^4	σ_h	S_5	S_5^2	S_5^3	S_5^4	$\omega = \exp(2\pi i/5)$		
A'	1	1	1	1	1	1	1	1	1	1		R_z	$x^2+y^2,\ z^2$
A''	1	1	1	1	1	-1	-1	-1	-1	-1	z		
$E_1'\ \Big\{$	1	ω	ω^2	ω^3	ω^4	1	ω	ω^2	ω^3	ω^4	$\Big\}\,x,y$		
	1	ω^*	ω^{*2}	ω^{*3}	ω^{*4}	1	ω^*	ω^{*2}	ω^{*3}	ω^{*4}			
$E_1''\ \Big\{$	1	ω	ω^2	ω^3	ω^4	-1	$-\omega$	$-\omega^2$	$-\omega^3$	$-\omega^4$		$\Big\}\,R_x,R_y$	$\Big\}\,yz,\ zx$
	1	ω^*	ω^{*2}	ω^{*3}	ω^{*4}	-1	$-\omega^*$	$-\omega^{*2}$	$-\omega^{*3}$	$-\omega^{*4}$			
$E_2'\ \Big\{$	1	ω^2	ω^4	ω	ω^3	1	ω^2	ω^4	ω	ω^3			$\Big\}\,x^2-y^2,\ xy$
	1	ω^{*2}	ω^{*4}	ω^*	ω^{*3}	1	ω^{*2}	ω^{*4}	ω^*	ω^{*3}			
$E_2''\ \Big\{$	1	ω^2	ω^4	ω	ω^3	-1	$-\omega^2$	$-\omega^4$	$-\omega$	$-\omega^3$			
	1	ω^{*2}	ω^{*4}	ω^*	ω^{*3}	-1	$-\omega^{*2}$	$-\omega^{*4}$	$-\omega^*$	$-\omega^{*3}$			

C_{6h}	E	C_6	C_3	C_2	C_3^2	C_6^5	i	S_3^2	S_6^5	σ_h	S_6	S_3	$\omega = \exp(2\pi i/6)$		
A_g	1	1	1	1	1	1	1	1	1	1	1	1		R_z	$x^2+y^2,\ z^2$
A_u	1	1	1	1	1	1	-1	-1	-1	-1	-1	-1	z		
B_g	1	-1	1	-1	1	-1	1	-1	1	-1	1	-1			
B_u	1	-1	1	-1	1	-1	-1	1	-1	1	-1	1			
$E_{1g}\ \Big\{$	1	ω	ω^2	-1	$-\omega$	$-\omega^2$	1	ω	ω^2	-1	$-\omega$	$-\omega^2$		$\Big\}\,R_x,R_y$	$\Big\}\,yz,\ zx$
	1	ω^*	ω^{*2}	-1	$-\omega^*$	$-\omega^{*2}$	1	ω^*	ω^{*2}	-1	$-\omega^*$	$-\omega^{*2}$			
$E_{1u}\ \Big\{$	1	ω	ω^2	-1	$-\omega$	$-\omega^2$	-1	$-\omega$	$-\omega^2$	1	ω	ω^2	$\Big\}\,x,y$		
	1	ω^*	ω^{*2}	-1	$-\omega^*$	$-\omega^{*2}$	-1	$-\omega^*$	$-\omega^{*2}$	1	ω^*	ω^{*2}			
$E_{2g}\ \Big\{$	1	ω^2	ω^4	1	ω^2	ω^4	1	ω^2	ω^4	1	ω^2	ω^4			$\Big\}\,x^2-y^2,\ xy$
	1	ω^{*2}	ω^{*4}	1	ω^{*2}	ω^{*4}	1	ω^{*2}	ω^{*4}	1	ω^{*2}	ω^{*4}			
$E_{2u}\ \Big\{$	1	ω^2	ω^4	1	ω^2	ω^4	-1	$-\omega^2$	$-\omega^4$	-1	$-\omega^2$	$-\omega^4$			
	1	ω^{*2}	ω^{*4}	1	ω^{*2}	ω^{*4}	-1	$-\omega^{*2}$	$-\omega^{*4}$	-1	$-\omega^{*2}$	$-\omega^{*4}$			

S_2	E	i	$S_2 = C_i$		
A_g	1	1		R_x, R_y, R_z	$x^2, y^2, z^2, xy, yz, zx$
A_u	1	-1	x, y, z		

S_4	E	S_4	C_2	S_4^3			
A	1	1	1	1		R_z	$x^2 + y^2, z^2$
B	1	-1	1	-1	z		$x^2 - y^2, xy$
E $\Big\{$	1	i	-1	$-i$	$\Big\}\, x, y$	$\Big\}\, R_x, R_y$	$\Big\}\, yz, zx$
	1	$-i$	-1	i			

S_6	E	C_3	C_3^2	i	S_6^5	S_6	$\omega = \exp(2\pi i/3)$		
A_g	1	1	1	1	1	1		R_z	$x^2 + y^2, z^2$
A_u	1	1	1	-1	-1	-1	z		
E_g $\Big\{$	1	ω	ω^2	1	ω	ω^2		$\Big\}\, R_x, R_y$	$\Big\}\, x^2 - y^2, xy, yz, zx$
	1	ω^*	ω^{*2}	1	ω^*	ω^{*2}			
E_u $\Big\{$	1	ω	ω^2	-1	$-\omega$	$-\omega^2$	$\Big\}\, x, y$		
	1	ω^*	ω^{*2}	-1	$-\omega^*$	$-\omega^{*2}$			

D_2	E	C_2 z	C_2 y	C_2 x	$D_2 = V$		
A	1	1	1	1			x^2, y^2, z^2
B_1	1	1	-1	-1	z	R_z	xy
B_2	1	-1	1	-1	y	R_y	zx
B_3	1	-1	-1	1	x	R_x	yz

D_3	E	$2C_3$	$3C_2$			
A_1	1	1	1			$x^2 + y^2, z^2$
A_2	1	1	-1	z	R_z	
E	2	-1	0	x, y	R_x, R_y	$x^2 - y^2, xy, yz, zx$

$\mathbf{D_4}$	E	$2C_4$	C_2	$2C_2'$	$2C_2''$			
A_1	1	1	1	1	1			$x^2 + y^2, z^2$
A_2	1	1	1	-1	-1	z	R_z	
B_1	1	-1	1	1	-1			
B_2	1	-1	1	-1	1			
E	2	0	-2	0	0	x, y	R_x, R_y	$x^2 - y^2, xy, yz, zx$

$\mathbf{D_5}$	E	$2C_5$	$2C_5^2$	$5C_2'$	$\varphi = 2\pi/5$		
A_1	1	1	1	1			$x^2 + y^2, z^2$
A_2	1	1	1	-1	z	R_z	
E_1	2	a	b	0	x, y	R_x, R_y	yz, zx
E_2	2	b	a	0			$x^2 - y^2, xy$

$$a = 2\cos\varphi = 2\cos 4\varphi = (\sqrt{5} - 1)/2 = 0.618034,$$
$$b = 2\cos 2\varphi = -(\sqrt{5} + 1)/2 = -1.618034.$$

$\mathbf{D_6}$	E	$2C_6$	$2C_3$	C_2	$3C_2'$	$3C_2''$			
A_1	1	1	1	1	1	1			$x^2 + y^2, z^2$
A_2	1	1	1	1	-1	-1		R_z	
B_1	1	-1	1	-1	1	-1			
B_2	1	-1	1	-1	-1	1	z		
E_1	2	1	-1	-2	0	0	x, y	R_x, R_y	yz, zx
E_2	2	-1	-1	2	0	0			$x^2 - y^2, xy$

$\mathbf{D_{2d}}$	E	$2S_4$	C_2	$2C_2'$	$2C_2''$	$\mathbf{D_{2d}} = \mathbf{V_d}$		
A_1	1	1	1	1	1			$x^2 + y^2, z^2$
A_2	1	1	1	-1	-1		R_z	
B_1	1	-1	1	1	-1			$x^2 - y^2$
B_2	1	-1	1	-1	1	z		xy
E	2	0	-2	0	0	x, y	R_x, R_y	yz, zx

$\mathbf{D}_{3d}$	E	$2C_3$	$3C_2'$	i	$2S_3$	$3\sigma_d$			
A_{1g}	1	1	1	1	1	1			$x^2+y^2,\ z^2$
A_{1u}	1	1	1	-1	-1	-1			
A_{2g}	1	1	-1	1	1	-1		R_z	
A_{2u}	1	1	-1	-1	-1	1	z		
E_g	2	-1	0	2	-1	0		$R_x,\ R_y$	$x^2-y^2,\ xy,\ yz,\ zx$
E_u	2	-1	0	-2	1	0	$x,\ y$		

$\mathbf{D}_{4d}$	E	$2S_8$	$2C_4$	$2S_8^3$	C_2	$4C_2'$	$4\sigma_d$			
A_1	1	1	1	1	1	1	1			$x^2+y^2,\ z^2$
A_2	1	1	1	1	1	-1	-1		R_z	
B_1	1	-1	1	-1	1	1	-1			
B_2	1	-1	1	-1	1	-1	1	z		
E_1	2	$\sqrt{2}$	0	$-\sqrt{2}$	-2	0	0	$x,\ y$		
E_2	2	0	-2	0	2	0	0			$x^2-y^2,\ xy$
E_3	2	$-\sqrt{2}$	0	$\sqrt{2}$	-2	0	0		$R_x,\ R_y$	$yz,\ zx$

$\mathbf{D}_{5d}$	E	$2C_5$	$2C_5^2$	$5C_2'$	i	$2S_{10}^3$	$2S_{10}$	$5\sigma_d$	$\varphi=2\pi/5$		
A_{1g}	1	1	1	1	1	1	1	1			$x^2+y^2,\ z^2$
A_{1u}	1	1	1	1	-1	-1	-1	-1			
A_{2g}	1	1	1	-1	1	1	1	-1		R_z	
A_{2u}	1	1	1	-1	-1	-1	-1	1	z		
E_{1g}	2	a	b	0	2	a	b	0		$R_x,\ R_y$	$yz,\ zx$
E_{1u}	2	a	b	0	-2	$-a$	$-b$	0	$x,\ y$		
E_{2g}	2	b	a	0	2	b	a	0			$x^2-y^2,\ xy$
E_{2u}	2	b	a	0	-2	$-b$	$-a$	0			

$$a = 2\cos\varphi = 2\cos 4\varphi = (\sqrt{5}-1)/2 = 0{,}618034,$$
$$b = 2\cos 2\varphi = -(\sqrt{5}+1)/2 = -1{,}618034\,.$$

$\mathbf{D}_{6d}$	E	$2S_{12}$	$2C_6$	$2S_4$	$2C_3$	$2S_{12}^5$	C_2	$6C_2'$	$6\sigma_d$			
A_1	1	1	1	1	1	1	1	1	1			x^2+y^2, z^2
A_2	1	1	1	1	1	1	1	-1	-1		R_z	
B_1	1	-1	1	-1	1	-1	1	1	-1			
B_2	1	-1	1	-1	1	-1	1	-1	1	z		
E_1	2	$\sqrt{3}$	1	0	-1	$-\sqrt{3}$	-2	0	0	x, y		
E_2	2	1	-1	-2	-1	1	2	0	0			x^2-y^2, xy
E_3	2	0	-2	0	2	0	-2	0	0			
E_4	2	-1	-1	2	-1	-1	2	0	0		R_x, R_y	yz, zx
E_5	2	$-\sqrt{3}$	1	0	-1	$\sqrt{3}$	-2	0	0			

$\mathbf{D}_{2h}$	E	C_2 z	C_2 y	C_2 x	i	σ xy	σ zx	σ yz	$\mathbf{D}_{2h} = \mathbf{V}_h$		
A_g	1	1	1	1	1	1	1	1			x^2, y^2, z^2
A_u	1	1	1	1	-1	-1	-1	-1			
B_{1g}	1	1	-1	-1	1	1	-1	-1		R_z	xy
B_{1u}	1	1	-1	-1	-1	-1	1	1	z		
B_{2g}	1	-1	1	-1	1	-1	1	-1		R_y	zx
B_{2u}	1	-1	1	-1	-1	1	-1	1	y		
B_{3g}	1	-1	-1	1	1	-1	-1	1		R_x	yz
B_{3u}	1	-1	-1	1	-1	1	1	-1	x		

$\mathbf{D}_{3h}$	E	$2C_3$	$3C_2'$	σ_h	$2S_3$	$3\sigma_v$			
A_1'	1	1	1	1	1	1			x^2+y^2, z^2
A_1''	1	1	1	-1	-1	-1			
A_2'	1	1	-1	1	1	-1		R_z	
A_2''	1	1	-1	-1	-1	1	z		
E'	2	-1	0	2	-1	0	x, y		x^2-y^2, xy
E''	2	-1	0	-2	1	0		R_x, R_y	yz, zx

D_{4h}	E	$2C_4$	C_2	$2C_2'$	$2C_2''$	i	$2S_4$	σ_h	$2\sigma_v$	$2\sigma_d$			
A_{1g}	1	1	1	1	1	1	1	1	1	1			x^2+y^2, z^2
A_{1u}	1	1	1	1	1	-1	-1	-1	-1	-1			
A_{2g}	1	1	1	-1	-1	1	1	1	-1	-1		R_z	
A_{2u}	1	1	1	-1	-1	-1	-1	-1	1	1	z		
B_{1g}	1	-1	1	1	-1	1	-1	1	1	-1			x^2-y^2
B_{1u}	1	-1	1	1	-1	-1	1	-1	-1	1			
B_{2g}	1	-1	1	-1	1	1	-1	1	-1	1			xy
B_{2u}	1	-1	1	-1	1	-1	1	-1	1	-1			
E_g	2	0	-2	0	0	2	0	-2	0	0		R_x, R_y	yz, zx
E_u	2	0	-2	0	0	-2	0	2	0	0	x, y		

D_{5h}	E	$2C_5$	$2C_5^2$	$5C_2'$	σ_h	$2S_5$	$2S_5^3$	$5\sigma_v$	$\varphi = 2\pi/5$			
A_1'	1	1	1	1	1	1	1	1				x^2+y^2, z^2
A_1''	1	1	1	1	-1	-1	-1	-1				
A_2'	1	1	1	-1	1	1	1	-1		R_z		
A_2''	1	1	1	-1	-1	-1	-1	1	z			
E_1'	2	a	b	0	2	a	b	0	x, y			
E_1''	2	a	b	0	-2	$-a$	$-b$	0		R_x, R_y		yz, zx
E_2'	2	b	a	0	2	b	a	0				x^2-y^2, xy
E_2''	2	b	a	0	-2	$-b$	$-a$	0				

$$a = 2\cos\varphi = 2\cos 4\varphi = (\sqrt{5}-1)/2 = 0{,}618\,034,$$
$$b = 2\cos 2\varphi = -(\sqrt{5}+1)/2 = -1{,}618\,034\,.$$

$\mathbf{D}_{6h}$	E	$2C_6$	$2C_3$	C_2	$3C_2'$	$3C_2''$	i	$2S_3$	$2S_6$	σ_h	$3\sigma_d$	$3\sigma_v$			
A_{1g}	1	1	1	1	1	1	1	1	1	1	1	1			$x^2 + y^2, z^2$
A_{1u}	1	1	1	1	1	1	-1	-1	-1	-1	-1	-1			
A_{2g}	1	1	1	1	-1	-1	1	1	1	1	-1	-1		R_z	
A_{2u}	1	1	1	1	-1	-1	-1	-1	-1	-1	1	1	z		
B_{1g}	1	-1	1	-1	1	-1	1	-1	1	-1	1	-1			
B_{1u}	1	-1	1	-1	1	-1	-1	1	-1	1	-1	1			
B_{2g}	1	-1	1	-1	-1	1	1	-1	1	-1	-1	1			
B_{2u}	1	-1	1	-1	-1	1	-1	1	-1	1	1	-1			
E_{1g}	2	1	-1	-2	0	0	2	1	-1	-2	0	0		R_x, R_y	yz, zx
E_{1u}	2	1	-1	-2	0	0	-2	-1	1	2	0	0	x, y		
E_{2g}	2	-1	-1	2	0	0	2	-1	-1	2	0	0			$x^2 - y^2, xy$
E_{2u}	2	-1	-1	2	0	0	-2	1	1	-2	0	0			

T	E	$4C_3$	$4C_3^2$	$3C_2$	$\omega = \exp(2\pi i/3)$		
A	1	1	1	1			$x^2 + y^2 + z^2$
$E\begin{cases}\\\\\end{cases}$	1	ω	ω^2	1			$\left.\begin{array}{l} x^2 + y^2 - 2z^2 \\ x^2 - y^2 \end{array}\right\}$
	1	ω^*	ω^{*2}	1			
F	3	0	0	-1	x, y, z	R_x, R_y, R_z	xy, yz, zx

T_d	E	$8C_3$	$3C_2$	$6S_4$	$6\sigma_d$		
A_1	1	1	1	1	1		$x^2 + y^2 + z^2$
A_2	1	1	1	-1	-1		
E	2	-1	2	0	0		$x^2 + y^2 - 2z^2,\ x^2 - y^2$
F_1	3	0	-1	1	-1	R_x, R_y, R_z	
F_2	3	0	-1	-1	1	x, y, z	xy, yz, zx

T_h	E	$4C_3$	$4C_3^2$	$3C_2$	i	$4S_6^5$	$4S_6$	$3\sigma_n$	$\omega = \exp(2\pi i/3)$		
A_g	1	1	1	1	1	1	1	1			$x^2 + y^2 + z^2$
A_u	1	1	1	1	-1	-1	-1	-1			
$E_g\begin{cases}\\\\\end{cases}$	1	ω	ω^2	1	1	ω	ω^2	1			$\left.\begin{array}{l} x^2 + y^2 - 2z^2 \\ x^2 - y^2 \end{array}\right\}$
	1	ω^*	ω^{*2}	1	1	ω^*	ω^{*2}	1			
$E_u\begin{cases}\\\\\end{cases}$	1	ω	ω^2	1	-1	$-\omega$	$-\omega^2$	-1			
	1	ω^*	ω^{*2}	1	-1	$-\omega^*$	$-\omega^{*2}$	-1			
F_g	3	0	0	-1	3	0	0	-1		R_x, R_y, R_z	xy, yz, zx
F_u	3	0	0	-1	-3	0	0	1	x, y, z		

O	E	$8C_3$	$3C_2$	$6C_4$	$6C_2'$		
A_1	1	1	1	1	1		$x^2 + y^2 + z^2$
A_2	1	1	1	-1	-1		
E	2	-1	2	0	0		$x^2 + y^2 - 2z^2,\ x^2 - y^2$
F_1	3	0	-1	1	-1	x, y, z	R_x, R_y, R_z
F_2	3	0	-1	-1	1		xy, yz, zx

I	E	$12C_5$	$12C_5^2$	$20C_3$	$15C_2$			
A	1	1	1	1	1	x, y, z	R_x, R_y, R_z	$x^2 + y^2 + z^2$
F_1	3	a	b	0	-1			
F_2	3	b	a	0	-1			
G	4	-1	-1	1	0			$\begin{cases} x^2 + y^2 - 2z^2 \\ x^2 - y^2, xy, yz, zx \end{cases}$
H	5	0	0	-1	1			

$a = (1 + \sqrt{5})/2 = 1{,}618034,$
$b = (1 - \sqrt{5})/2 = -0{,}618034.$

$\mathbf{C}_{\infty v}$	E	$2C_\infty^\varphi$		C_2	$\infty\sigma_v$	$0 < \varphi < \pi$		
$A_1 = \Sigma^+$	1	1	...	1	1	z		$x^2 + y^2, z^2$
$A_2 = \Sigma^-$	1	1	...	1	-1		R_z	
$E_1 = \Pi$	2	$2\cos\varphi$	...	-2	0	x, y	R_x, R_y	yz, zx
$E_2 = \Delta$	2	$2\cos 2\varphi$	...	2	0			$x^2 - y^2, xy$
$E_3 = \Phi$	2	$2\cos 3\varphi$	...	-2	0			
...	...	...	...	...	...			

The running header on this page reads, in the right margin: "Appendix 5 Irreducible Representations of Symmetry Groups", with the page number 197 at the bottom of that margin.

$D_{\infty h}$	E	$2C_\infty^\varphi$	...	C_2	$\infty\sigma_v$	i	$2S_\infty^\varphi$	...	σ_h	$\infty C_2'$	$0 < \varphi < \pi$		
Σ_g^+	1	1	...	1	1	1	1	...	1	1			$x^2 - y^2,\ z^2$
Σ_u^+	1	1	...	1	1	-1	-1	...	-1	-1	z		
Σ_g^-	1	1	...	1	-1	1	1	...	1	-1		R_z	
Σ_u^-	1	1	...	1	-1	-1	-1	...	-1	1			
Π_g	2	$2\cos\varphi$	...	-2	0	2	$2\cos\varphi$	...	-2	0		$R_x,\ R_y$	$yz,\ zx$
Π_u	2	$2\cos\varphi$	...	-2	0	-2	$-2\cos\varphi$	...	2	0	$x,\ y$		
Δ_g	2	$2\cos 2\varphi$	...	2	0	2	$2\cos 2\varphi$	...	2	0			$x^2 - y^2,\ xy$
Δ_u	2	$2\cos 2\varphi$	...	2	0	-2	$-2\cos 2\varphi$	...	-2	0			
.	.	.	...	.	.	.	.	...	.	.			

Symmetric groups, S_n, $n \leq 6$

S_1	[1]
	1
Γ_0	1

S_2	$[1]^2$	[2]
	1	1
Γ_0	1	1
Γ_1	1	-1

S_3	$[1]^3$	[3]	[1][2]
	1	2	3
Γ_0	1	1	1
Γ_1	1	1	-1
Γ_2	2	-1	0

S_4	$[1]^4$	[1][3]	$[2]^2$	$[1]^2[2]$	[4]
	1	8	3	6	6
Γ_0	1	1	1	1	1
Γ_1	1	1	1	-1	-1
Γ_2	2	-1	2	0	0
Γ_3	3	0	-1	1	-1
Γ_4	3	0	-1	-1	1

S_5	$[1]^5$	[5]	$[1]^2[3]$	$[1][2]^2$	$[1]^3[2]$	[2][3]	[1][4]
	1	24	20	15	10	20	30
Γ_0	1	1	1	1	1	1	1
Γ_1	1	1	1	1	-1	-1	-1
Γ_2	4	-1	1	0	2	-1	0
Γ_3	4	-1	1	0	-2	1	0
Γ_4	5	0	-1	1	1	1	-1
Γ_5	5	0	-1	1	-1	-1	1
Γ_6	6	1	0	-2	0	0	0

S_6	$[1]^6$	$[1][5]$	$[1]^3[3]$	$[1]^2[2]^2$	$[3]^2$	$[2][4]$	$[1]^4[2]$	$[1]^2[4]$	$[6]$	$[2]^3$	$[1][2][3]$
	1	144	40	45	40	90	15	90	120	15	120
Γ_0	1	1	1	1	1	1	1	1	1	1	1
Γ_1	1	1	1	1	1	1	−1	−1	−1	−1	−1
Γ_2	5	0	2	1	−1	−1	3	1	−1	−1	0
Γ_3	5	0	2	1	−1	−1	−3	−1	1	1	0
Γ_4	5	0	−1	1	2	−1	1	−1	0	−3	1
Γ_5	5	0	−1	1	2	−1	−1	1	0	3	−1
Γ_6	9	−1	0	1	0	1	3	−1	0	3	0
Γ_7	9	−1	0	1	0	1	−3	1	0	−3	0
Γ_8	10	0	1	−2	1	0	2	0	1	−2	−1
Γ_9	10	0	1	−2	1	0	−2	0	−1	2	1
Γ_{10}	16	1	−2	0	−2	0	0	0	0	0	0

Appendix 6

The Symbols Used

In the right-hand side column of the following list the chapter, section, paragraph or equation are indicated, where the respective notion has been defined.

A	TEMO subunit	13.1
A, $A(G)$	adjacency matrix (of the graph G)	4.3.1
$A(G)$	automorphism group (of the graph G)	9.2
A, A_1, A', A_g, A_{1g}	totally symmetric irreducible representation	7.2 and 8.1
$\mathbf{A}_n$	alternant group	9.5
a_k	coefficient of $\varphi(G)$	Eq. (4.21)
a_{ij}	element of the adjacency matrix	Eq. (4.12)
B	TEMO subunit	13.1
B	one-dimensional irreducible representation	8.1
$b(G, k)$	coefficient of $\varphi(G)$ of bipartite graphs	Eq. (6.56)
C_n	cycle with n vertices	4.1.6
C_n	n-fold rotation axis	8.1
C_i	i-th eigenvector of the graph	Eq. (4.15)
$\mathbf{C}_i$	symmetry group; $\mathbf{C}_i \simeq \mathbf{S}_2$	8.2.1
$\mathbf{C}_n$, $\mathbf{C}_{nv}$, $\mathbf{C}_{nh}$	symmetry groups	8.2.1
$\mathbf{C}_n$	cyclic group	9.5
$\mathbf{C}_{\infty v}$	symmetry group	8.2.3
$c(S)$	number of cycles in the SACHS graph S	4.2.1
$cs(P)$	cycle structure of the permutation P	9.3
c_{ij}	LCAO coefficient	Eq. (5.1)
$D(G)$	distance matrix of the graph G	4.1.4
D_v	distance vector of the vertex v	11.1.1
$\mathbf{D}_n$, $\mathbf{D}_{nv}$, $\mathbf{D}_{nh}$	symmetry groups	8.2.1
$\mathbf{D}_{\infty h}$	symmetry group	8.2.3
$\mathbf{D}_n$	dihedral group	9.5
$d(u, v)$	distance between the vertices u and v	4.1.4
$d(v)$, $d(v, G)$	distance number of the vertex v	11.1.1
d_{ij}	element of the distance matrix	Eq. (11.2)
E	identity group element	7
E, E_π	total π-electron energy	Eqs. (12.2) and (12.4)
E_i	energy of the i-th MO	Eq. (5.1)
$\mathscr{E}$, $\mathscr{E}(G)$	edge set (of the graph G)	4.1.4
E	two-dimensional irreducible representation	8.1
$\mathbf{E}_n$	identic group of degree n	9.5
e, e_i	edge	4.1.2 and 4.1.3
G	graph	4.1.2
$\mathbf{G}$	group	7
$\mathscr{G}$, $\mathscr{G}_n$	set of all graphs, set of all graphs with n vertices	4.1.2
$\mathscr{G}_{a,b}$	set of all bipartite graphs with $a + b$ vertices, $a \leqq b$	6.3.1

$g(\mathbf{G})$	degree of the automorphism group $\mathbf{G}$	9.2
g_v	degree of the vertex v	4.1
$\mathscr{H}$	Hamiltonian operator	5 and 8.4.6
$\mathbf{H}$	Hamiltonian matrix	5
h	number of hexagons in a benzenoid graph	6.4.1
$h, h(\mathbf{G})$	order (of the group $\mathbf{G}$)	7.2
$\mathbf{I}, \mathbf{I}_n$	unit matrix (of the order n)	Appendix 1
$\mathbf{I}, \mathbf{I}_h$	icosahedral groups	8.2.2
i	inversion	8.1
K	number of KEKULÉ structures	6.4.2
K_n	complete graph with n vertices	4.1.6
$K_{a,b}$	complete bipartite graph with $a + b$ vertices, $a \leq b$	6.3.1
$K_{1,n-1}$	star with n vertices	6.1.2
$k, k(G)$	number of components (of the graph G)	4.1.4
$L(G)$	line graph of the graph G	4.4
l_i	dimension of the i-th irreducible representation	7.3
m	number of edges; $m = |\mathscr{E}|$	4.1.4
$m(G, k)$	number of k-matchings of the graph G	4.2.2
n	number of vertices; $n = |\mathscr{V}|$	4.1.4
n_i	number of internal vertices of a benzenoid graph	6.4.1
$\mathbf{O}, \mathbf{O}_n$	zero matrix (of order n)	Appendix 1
$\mathbf{O}, \mathbf{O}_h$	octahedral groups	8.2.2
P_n	path with n vertices	6.1.2
P_{uv}	path between the vertices u and v	13.2
R	element of a symmetry group, symmetry operation	7.3 and 8
R_x, R_y, R_z	rotation modes	Eq. (8.27)
S	SACHS graph	4.2.1
S	topomer in TEMO	13.1
S_n	n-fold rotation-reflection axis	8.1
$\mathscr{S}_i, \mathscr{S}_i(G)$	set of SACHS graphs with i vertices (contained in the graph G)	4.2.1
$\mathbf{S}_n$	symmetry group	8.2.3
$\mathbf{S}_n$	symmetric group	9.5
T	tree	6.1.1
T	topomer in TEMO	13.1
T_x, T_y, T_z	translation modes	Eq. (8.26)
$\mathscr{T}_n$	set of all trees with n vertices	6.1.1
$\mathbf{T}, \mathbf{T}_d, \mathbf{T}_h$	tetrahedral groups	8.2.2
u	vertex	4.1.4
$\mathscr{V}, \mathscr{V}(G)$	vertex set (of the graph G)	4.1.4
v, v_i	vertex	4.1.2
$W, W(G)$	WIENER index (of the graph G)	Eq. (11.4)
W	weight matrix	6.5.2
Z	cycle	4.3.3
$Z, Z(G)$	HOSOYA's index (of the graph G)	Eq. (11.14)
α	HMO parameter	5 and 13.4
$\alpha(G, x)$	matching polynomial	Eq. (4.11)
β	HMO parameter	5 and 13.4
Γ_i	representation of a group	7.2
$\Delta(x)$	$\varphi(T) - \varphi(S)$ in TEMO	Eq. (13.2)
δ_{ij}	KRONECKER's symbol; $\delta_{ii} = 1$ and $\delta_{ij} = 0$ if $i \neq j$	
η_j	number of electrons in the j-th MO	12.1
$\lambda_j, \lambda_j(G)$	eigenvalue (of the graph G)	Eq. (4.15)

$\mu(G, x)$	μ-polynomial	Eq. (4.36)
$\sigma_h, \sigma_v, \sigma_d$	reflections	8.1
$\varphi(G, x)$	characteristic polynomial	Eq. (4.18)
$\chi(R), \chi_i(R)$	character of the group element R	
	(in the i-th irreducible representation)	Eq. (7.22)
ψ_i	wave function, molecular orbital	Eqs. (5.2) and (8.37)
$\overline{G}$	complement of the graph G	4.4
M^{-1}	inverse of the matrix M;	
	$M^{-1}M = MM^{-1} = I$	Appendix 1
R^{-1}	inverse of the group element R;	
	$R^{-1}R = RR^{-1} = E$	7
$G_1 \cup G_2$	union of the graphs G_1 and G_2	4.4
$G_1 \oplus G_2$	compound of the graphs G_1 and G_2	4.4
$G_1[G_2]$	composition of the graphs G_1 and G_2	4.4
$\mathbf{G}_1 \odot \mathbf{G}_2$	product (either direct or semidirect) of the	
	groups $\mathbf{G}_1$ and $\mathbf{G}_2$	7.5
$\mathbf{G}_1 \oplus \mathbf{G}_2$	direct product of the groups $\mathbf{G}_1$ and $\mathbf{G}_2$	7.5
$\mathbf{G}_1 \oslash \mathbf{G}_2$	semidirect product of the groups $\mathbf{G}_1$ and $\mathbf{G}_2$	7.5
$\mathbf{G}_1[\mathbf{G}_2]$	wreath product of the groups $\mathbf{G}_1$ and $\mathbf{G}_2$	9.6
$G_1 \simeq G_2$	the graphs G_1 and G_2 are isomorphic	4.4
$G_1 \succ G_2$	$m(G_1, k) \geqq m(G_2, k)$ for all k	6.1.4
$G_1 > \cdot\, G_2$	$b(G_1, k) \geqq b(G_2, k)$ for all k	12.5
$\mathbf{G}_1 \simeq \mathbf{G}_2$	the groups $\mathbf{G}_1$ and $\mathbf{G}_2$ are isomorphic	7.7
$\mathbf{G}_1 \subset \mathbf{G}_2$	$\mathbf{G}_1$ is a proper subgroup of the group $\mathbf{G}_2$	7.5
$\emptyset$	empty set	
$|\mathscr{X}|$	number of elements of the set $\mathscr{X}$	
$x \in \mathscr{X}$	x is an element of the set $\mathscr{X}$	
$\mathscr{X} \cup \mathscr{Y}$	union of the sets $\mathscr{X}$ and $\mathscr{Y}$	
$\mathscr{X} \cap \mathscr{Y}$	intersection of the sets $\mathscr{X}$ and $\mathscr{Y}$	
$\mathscr{X} \subseteq \mathscr{Y}$	$\mathscr{X}$ is a subset of the set $\mathscr{Y}$	
$\mathscr{X} \otimes \mathscr{Y}$	cartesian product of the sets $\mathscr{X}$ and $\mathscr{Y}$	

Literature

Bibliography

1. Altman, S. L.: Induced representations in crystals and molecules. London: Academic Press 1977
2. Anh, N. T.: Les règles de Woodward-Hoffmann. Paris: Ediscience 1970; Anh, N. T.: Die Woodward-Hoffmann Regeln und ihre Anwendungen. Weinheim: Verlag Chemie 1972
3. Au-chin, T., Yuan-sun, K., Shu-sun, D., Guo-sun, Y.: Graph theory and molecular orbitals. Peking: Science Press 1980 (in Chinese)
4. Bader, R. F. W.: A theory of molecular structure, in: King, R. B. (Ed.): Chemical applications of topology and graph theory. Amsterdam: Elsevier 1983, pp. 40–56
5. Badger, G. M.: Aromatic character and aromaticity. Cambridge: Cambridge University Press 1969
6. Balaban, A. T. (Ed.): Chemical applications of graph theory. London: Academic Press 1976
7. Balaban, A. T., Motoc, I., Bonchev, D., Mekenyan, O.: Topological indices for structure-activity correlations. In: Topics in Current Chemistry, Vol. 114, pp. 21–55. Berlin Heidelberg New York Tokyo, Springer 1983
8. Berge, C.: The theory of graphs and its application. London: Methuen 1962
9. Bonchev, D.: Information theoretic indices for characterization of chemical structures. Chichester: Research Studies Press 1983
10. Busacker, R. G., Saaty, T.: Finite graphs and networks. New York: McGraw-Hill 1965
11. Chisholm, C. D. H.: Group theoretical techniques in quantum chemistry. London: Academic Press 1976
12. Clar, E.: Polycyclic hydrocarbons. (2 Volumes). London: Academic Press 1964
13. Clar, E.: The aromatic sextet. London: Wiley 1972
14. Cotton, F. A.: Chemical applications of group theory. New York: Wiley-Interscience 1971
15. Coulson, C. A., Streitwieser Jr., A.: Dictionary of π-electron calculations. San Francisco: Freeman 1965
16. Cvetković, D. M., Doob, M., Sachs, H.: Spectra of graphs — theory and application. New York: Academic Press 1980
17. Dewar, M. J. S., Dougherty, R. C.: The PMO theory of organic chemistry. New York: Plenum Press 1975
18. Dmitriev, I. S.: Molecules without chemical bonds. Moscow: Mir 1980 (in English)
19. Dugundji, J.: Topology. Boston: Allyn and Bacon 1966
20. Eyring, H., Walter, J., Kimball, G. E.: Quantum chemistry. New York: Wiley 1961
21. Fackler Jr., J. P.: Symmetry in coordination chemistry. New York: Academic Press 1971
22. Frisch, H. L., Wasserman, E.: Chemical topology. J. Amer. Chem. Soc. *83*, 3789–3795 (1961)
23. Godsil, C. D., Gutman, I.: On the theory of the matching polynomial. J. Graph Theory *5*, 137–144 (1981)
24. Graovac, A., Gutman, I., Trinajstić, N.: Topological approach to the chemistry of conjugated molecules. Berlin Heidelberg New York: Springer 1977
25. Gutman, I.: The energy of a graph. Berichte Math.-Statist. Sekt. Forschungszentrum Graz *103*, 1–22 (1978)

26. Gutman, I.: Topology and stability of conjugated hydrocarbons. The dependence of total π-electron energy on molecular topology. Bull. Soc. Chim. Beograd *43*, 761–774 (1978) (in Serbo-Croatian)
27. Gutman, I.: The matching polynomial. Match (Math. Chem.) *6*, 75–91 (1979)
28. Gutman, I.: Topological properties of benzenoid molecules. Bull. Soc. Chim. Beograd *47*, 453–471 (1982)
29. Gutman, I.: Topological properties of benzenoid systems. XXI. Theorems, conjectures, unsolved problems. Croat. Chem. Acta *56*, 365–374 (1983)
30. Gutman, I., Trinajstić, N.: Graph theory and molecular orbitals. In: Topics in Current Chemistry, Vol. 42, pp. 49–93. Berlin Heidelberg New York: Springer 1973
31. Gutman, I., Trinajstić, N.: Graph spectral theory of conjugated molecules. Croat. Chem. Acta *47*, 507–533 (1975)
32. Hall, L. H.: Group theory and symmetry in chemistry. New York: McGraw-Hill 1969
33. Hamermesh, M.: Group theory and its application to physical problems. Reading (Mass.): Addison-Wesley 1964
34. Harary, F.: Graph theory. Reading (Mass.): Addison-Wesley 1972; Harary, F.: Graphentheorie. München: Oldenbourg 1974
35. Heilbronner, E., Bock, H.: Das HMO-Modell und seine Anwendung. Weinheim: Verlag Chemie 1968–1970
36. Heine, V.: Group theory in quantum mechanics. Oxford: Pergamon Press 1960
37. Herzberg, G.: Molecular spectra and molecular structure, Vol. 2. Infrared and Raman spectra of polyatomic molecules. New York: Van Nostrand 1945
38. Hinze, J. (Ed.): The permutation group in physics and chemistry. Berlin Heidelberg New York: Springer 1979
39. Jacimirskij, K. B.: Anwendung der Graphenmethode in der Chemie. Z. Chemie *13*, 201–213 (1973)
40. Jackman, L. M., Sternhell, S.: Applications of nuclear magnetic resonance spectroscopy in organic chemistry. Oxford: Pergamon Press 1969
41. James, G., Kerber, A.: The representation theory of the symmetric group. Encyclopedia of Mathematics and its Applications, Vol. 16. Reading (Mass.): Addison-Wesley 1981
42. Jansen, L., Boon, M.: Theory of finite groups. Amsterdam: North-Holland 1967
43. Jørgensen, C. K.: Absorption spectra and chemical bonding in complexes. Oxford: Pergamon Press 1962
44. Jotham, R. W.: Chemistry — a topological subject. Chem. Soc. Revs. *2*, 457–474 (1973)
45. Judd, B. R.: Operator techniques in atomic spectroscopy. New York: McGraw-Hill 1963
46. Kerber, A.: Representations of permutation groups I–II (Lecture Notes in Mathematics 240, 495). Berlin Heidelberg New York: Springer 1971, 1975
47. Kier, L. B., Hall, L. H.: Molecular connectivity in chemistry and drug research. New York: Academic Press 1976
48. Koster, G. F., Dimmock, J. O., Wheeler, R. G., Statz, H.: Properties of the thirty-two point groups. Cambridge: MIT Press 1963
49. Littlewood, D. E.: The theory of group characters. Oxford: Clarendon Press 1940
50. Matossi, F.: Gruppentheorie der Eigenschwingungen von Punktsystemen. Berlin Göttingen Heidelberg: Springer 1961
51. McWeeny, R.: Coulson's valence, 3. Ed. Oxford: Oxford University Press 1979; McWeeny, R.: Coulsons Chemische Bindung. Stuttgart: Hirzel 1984
52. Mead, C. A.: Symmetry and chirality. Topics Current Chemistry, Vol. 49. Berlin Heidelberg New York: Springer 1974
53. Mezey, P. G.: Reaction topology: Manifold theory of potential surfaces and quantum chemical synthesis design, in: King, R. B. (Ed.): Chemical applications of topology and graph theory. Amsterdam: Elsevier 1983, pp. 75–98
54. Nourse, J. G.: Application of the permutation group in dynamic stereochemistry, in: Hinze, J. (Ed.): The permutation group in physics and chemistry. Berlin Heidelberg New York: Springer 1979
55. Ore, O.: Theory of graphs. Providence: Amer. Math. Soc. 1973
56. Orgel, L. E.: An introduction to transition-metal chemistry. Ligand field theory. London: Methuen 1962

57. Sachs, H.: Einführung in die Theorie der endlichen Graphen. München: Hanser 1971
58. Salthouse, J. A., Ware, M. J.: Point group character tables and related data. Cambridge: University Press 1972
59. Sedláček, J.: Einführung in die Graphentheorie. Frankfurt: Deutsch-Verlag 1972
60. Schill, G.: Catenanes, rotaxanes and knots. New York: Academic Press 1971
61. Schuster, P.: Ligandenfeldtheorie. Weinheim: Verlag Chemie 1973
62. Streitwieser Jr., A.: Molecular orbital theory for organic chemists. New York: Wiley 1961
63. Streitwieser Jr., A., Brauman, J. I.: Supplemental tables of molecular orbital calculations. (2 Volumes). Oxford: Pergamon Press 1965
64. Trinajstić, N.: Computing the characteristic polynomial of a conjugated system using the Sachs theorem. Croat. Chem. Acta *49*, 593–633 (1977)
65. Trinajstić, N.: Chemical graph theory. (2 Volumes.) Boca Raton: CRC Press 1983
66. Wilson Jr., E. B., Decius, J. C., Cross, P. C.: Molecular vibrations. New York: McGraw-Hill 1955
67. Wilson, R. J.: Introduction to graph theory. Edinburgh: Oliver and Boyd 1972
68. Woodward, R. B., Hoffmann, R.: The conservation of orbital symmetry. New York: Academic Press 1969

References

69. Aihara, J.: J. Amer. Chem. Soc. *98*, 2750 (1976)
70. Aihara, J.: J. Amer. Chem. Soc. *98*, 6840 (1976)
71. Altmann, S. L.: Proc. Roy. Soc. *A 298*, 184 (1967)
72. Balaban, A. T.: Tetrahedron *25*, 2949 (1969)
73. Balaban, A. T.: Pure Appl. Chem. *54*, 1075 (1982)
74. Balasubramanian, K., Randić, M.: Theoret. Chim. Acta *61*, 307 (1982)
75. Bonchev, D., Kamenski, D., Temkin, O. N.: J. Comput. Chem. *3*, 95 (1982)
76. Boschi, R., Clar, E., Schmidt, W.: J. Chem. Phys. *60*, 4406 (1974)
77. Clarke, B. L.: J. Chem. Phys. *64*, 4165 (1976)
78. Coulson, C. A.: Proc. Cambridge Phil. Soc. *36*, 201 (1940)
79. Coulson, C. A., Longuet-Higgins, H. C.: Proc. Roy. Soc. *A 192*, 16 (1948)
80. Coulson, C. A., Rushbrooke, G. S.: Proc. Cambridge Phil. Soc. *36*, 193 (1940)
81. Cvetković, D., Gutman, I., Trinajstić, N.: J. Chem. Phys. *61*, 2700 (1974)
82. de Moor, J. E., van der Kelen, G. P.: J. Organomet. Chem. *6*, 235 (1966)
83. Dewar, M. J. S., de Llano, C.: J. Amer. Chem. Soc. *91*, 789 (1969)
84. Dewar, M. J. S., Longuet-Higgins, H. C.: Proc. Roy. Soc. *A 214*, 482 (1952)
85. Dias, J. R.: Match (Math. Chem.) *17*, 175 (1985)
86. El-Basil, S.: Theoret. Chim. Acta *65*, 199 (1984)
87. El-Basil, S., Shalabi, A. S.: Match (Math. Chem.) *17*, 11 (1985)
88. Fabian, W., Motoc, I., Polansky, O. E.: Z. Naturforsch. *38a*, 916 (1983)
89. Flurry Jr., R. L.: Int. J. Quantum Chem. *25*, 309 (1984)
90. George, P.: Chem. Rev. *75*, 85 (1975)
91. Good, C. D., Ritter, D. M.: J. Amer. Chem. Soc. *84*, 1162 (1962)
92. Graovac, A., Gutman, I., Polansky, O. E.: Mh. Chemie *115*, 1 (1984)
93. Graovac, A., Gutman, I., Polansky, O. E.: J. Chem. Soc., Faraday Trans. II, *81*, 1543 (1985)
94. Graovac, A., Gutman, I., Trinajstić, N., Živković, T.: Theoret. Chim. Acta *26*, 67 (1972)
95. Graovac, A., Polansky, O. E.: Croat. Chem. Acta *57*, 1595 (1984)
96. Graovac, A., Polansky, O. E., Trinajstić, N., Tyutyulkov, N. N.: Z. Naturforsch. *30a*, 1696 (1975)
97. Gutman, I.: Croat. Chem. Acta *46*, 209 (1974)
98. Gutman, I.: Chem. Phys. Letters *24*, 283 (1974)
99. Gutman, I.: J. Chem. Phys. *66*, 1652 (1977)
100. Gutman, I.: Theoret. Chim. Acta *45*, 79 (1977)
101. Gutman, I.: Theoret. Chim. Acta *45*, 309 (1977)
102. Gutman, I.: Croat. Chem. Acta *51*, 299 (1978)
103. Gutman, I.: Match (Math. Chem.) *4*, 195 (1978)

104. Gutman, I.: Publ. Inst. Math. (Beograd) 27, 67 (1980); correction: ibid. 32, 61 (1982)
105. Gutman, I.: Z. Naturforsch. 36a, 1112 (1981)
106. Gutman, I.: Z. Naturforsch. 37a, 69 (1982)
107. Gutman, I.: J. Chem. Soc., Faraday Trans. II, 79, 337 (1983)
108. Gutman, I.: Match (Math. Chem.) 14, 71 (1983)
109. Gutman, I.: Z. Naturforsch. 39a, 152 (1984)
110. Gutman, I.: Theoret. Chim. Acta 66, 43 (1984)
111. Gutman, I.: Z. Phys. Chem. (Leipzig) 266, 59 (1985)
112. Gutman, I.: Chem. Phys. Letters 117, 614 (1985)
113. Gutman, I., Cvetković, D.: Publ. Inst. Math. (Beograd) 35, 33 (1984)
114. Gutman, I., Graovac, A., Polansky, O. E.: Chem. Phys. Letters 116, 206 (1985)
115. Gutman, I., Milun, M., Trinajstić, N.: J. Chem. Phys. 59, 2772 (1973)
116. Gutman, I., Milun, M., Trinajstić, N.: J. Amer. Chem. Soc. 99, 1692 (1977)
117. Gutman, I., Nedeljković, Lj., Teodorović, A. V.: Bull. Soc. Chim. Beograd 48, 495 (1983)
118. Gutman, I., Petrović, S.: Chem. Phys. Letters 97, 292 (1983)
119. Gutman, I., Polansky, O. E.: Theoret. Chim. Acta 60, 203 (1981)
120. Gutman, I., Teodorović, A. V., Bugarčić, Ž.: Bull. Soc. Chim. Beograd 49, 521 (1984)
121. Gutman, I., Teodorović, A. V., Nedeljković, Lj.: Theoret. Chim. Acta 65, 23 (1984)
122. Gutman, I., Zhang, F.: to be published
123. Hall, G. G.: Proc. Roy. Soc. A 229, 251 (1955)
124. Hall, G. G.: Bull. Inst. Math. Appl. 17, 70 (1981)
125. Harary, F., Harborth, H.: J. Comb. Inf. System Sci. 1, 1 (1976)
126. Heilbronner, E.: Helv. Chim. Acta 36, 170 (1953)
127. Hess Jr., B. A., Schaad, L. J.: J. Amer. Chem. Soc. 93, 305 (1971)
128. Hess Jr., B. A., Schaad, L. J.: J. Chem. Educ. 51, 640 (1974)
129. Hosoya, H.: Bull. Chem. Soc. Japan 44, 2332 (1971)
130. Hosoya, H.: J. Chem. Docum. 12, 181 (1972)
131. Hosoya, H., Kawasaki, K., Mizutani, K.: Bull. Chem. Soc. Japan 45, 3415 (1972)
132. Hosoya, H., Yamaguchi, T.: Tetrahedron Letters 4659 (1975)
133. Ichikawa, H., Ebisawa, Y.: J. Amer. Chem. Soc. 107, 1161 (1985)
134. Jiang, Y., Tang, A., Hoffmann, R.: Theoret. Chim. Acta 66, 183 (1984)
135. Joela, H.: Theoret. Chim. Acta 39, 241 (1975)
136. Kaulgud, M. V., Chitgopkar, V. H.: J. Chem. Soc., Faraday Trans. II, 73, 1385 (1977)
137. Koopmans, T.: Physica 1, 104 (1934)
138. Kruszewski, J., Polansky, O. E.: Match (Math. Chem.) 19, 243, 267 (1986)
139. Lall, R. S.: Match (Math. Chem.) 13, 17 (1982)
140. Lennard-Jones, J. E.: Proc. Roy. Soc. A 158, 280 (1937)
141. Liu, R. S. H.: J. Amer. Chem. Soc. 89, 112 (1967)
142. Longuet-Higgins, H. C.: Molec. Phys. 6, 445 (1963)
143. March, N. H.: J. Chem. Phys. 67, 4618 (1977)
144. McClelland, B. J.: J. Chem. Phys. 54, 640 (1971)
145. Merrifield, R. E., Simmons, H. E.: Theoret. Chim. Acta 55, 55 (1980)
146. Milun, M., Sobotka, Ž., Trinajstić, N.: J. Org. Chem. 37, 139 (1972)
147. Motoc, I., Balaban, A. T.: Rev. Roum. Chim. 26, 593 (1981)
148. Motoc, I., Polansky, O.E.: Z. Naturforsch. 39b, 1053 (1984)
149. Motoc, I., Silverman, J. N., Polansky, O. E.: Phys. Rev. A 28, 3673 (1983)
150. Motoc, I., Silverman, J. N., Polansky, O. E.: Chem. Phys. Letters 103, 285 (1984)
151. Motoc, I., Silverman, J. N., Polansky, O. E., Olbrich, G.: Theoret. Chim. Acta 67, 63 (1985)
152. Narumi, H., Hosoya, H.: Bull. Chem. Soc. Japan 53, 1228 (1980)
153. Polansky, O. E.: unpublished results
154. Polansky, O. E.: Match (Math. Chem.) 1, 183 (1975)
155. Polansky, O. E.: J. Mol. Struct. 113, 281 (1984)
156. Polansky, O. E.: Match (Math. Chem.) 18, 111 (1985)
157. Polansky, O. E.: Match (Math. Chem.) 18, 167 (1985)
158. Polansky, O. E., Bonchev, D.: to be published
159. Polansky, O. E., Mark, G.: Match (Math. Chem.) 18, 249 (1985)

160. Polansky, O. E., Rouvray, D. H.: Match (Math. Chem.) *2*, 63 (1976)
161. Polansky, O. E., Tyutyulkov, N. N.: Match (Math. Chem.) *3*, 149 (1977)
162. Polansky, O. E., Zander, M.: J. Mol. Struct. *84*, 361 (1982)
163. Polansky, O. E., Zander, M., Motoc, I.: Z. Naturforsch. *38a*, 196 (1983)
164. Pólya, G.: Acta Mathematica *68*, 145 (1937)
165. Prelog, V.: J. Mol. Catalysis *1*, 159 (1975/76)
166. Rigby, M. J., Mallion, R. B., Day, A. C.: Chem. Phys. Letters *51*, 178 (1977); correction: ibid. *53*, 418 (1978)
167. Ruch, E.: Acc. Chem. Res. *5*, 49 (1972)
168. Ruch, E., Schönhofer, A.: Theoret. Chim. Acta *19*, 225 (1970)
169. Ruedenberg, K.: J. Chem. Phys. *22*, 1878 (1954)
170. Ruedenberg, K.: J. Chem. Phys. *66*, 375 (1977)
171. Sachs, H.: Publ. Math. (Debrecen) *11*, 119 (1963)
172. Serre, J.: Adv. Quantum Chem. *8*, 1 (1974)
173. Schaad, L. J., Hess Jr., B. A.: J. Amer. Chem. Soc. *94*, 3068 (1972)
174. Schaad, L. J., Robinson, B. H., Hess Jr., B. A.: J. Chem. Phys. *67*, 4616 (1977)
175. Schmidt, W.: J. Chem. Phys. *66*, 828 (1977)
176. Schuster, P.: Theoret. Chim. Acta *3*, 278 (1965)
177. Türker, L.: Match (Math. Chem.) *16*, 83 (1984)
178. Walba, D. M., Richards, R. M., Haltiwanger, R. C.: J. Amer. Chem. Soc. *104*, 3219 (1982)
179. Wiener, H.: J. Amer. Chem. Soc. *60*, 17 (1947)
180. Wild, U., Keller, J., Günthard, H. H.: Theoret. Chim. Acta *14*, 383 (1969)
181. Woolley, R. G.: J. Amer. Chem. Soc. *100*, 1073 (1978)
182. Zander, M., Polansky, O. E.: Naturwiss. *71*, 623 (1984)
183. Zhang, F.: Kexue Tongbao *28*, 726 (1983)

Added in proof: Since the time the manuscript of this book has been given to the publisher, many relevant papers and books have appeared. Among them we wish to draw the reader's attention to the following ones:

a. Balasubramanian, K.: Application of combinatorics and graph theory to spectroscopy and quantum chemistry. Chem. Rev. *85*, 599–618 (1985)
b. Trinajstić, N. (Ed.): Mathematics and computational concepts in chemistry. Chichester: Ellis Horwood 1986
c. Wald, D.: Gruppentheorie für Chemiker. Weinheim: Verlag Chemie 1985

Subject Index

Lecture Notes in Chemistry

Editors: G. Berthier, M. J. S. Dewar, H. Fischer, K. Fukui, G. G. Hall. J. Hinze, H. H. Jaffé, J. Jortner, W. Kutzelnigg, K. Ruedenberg, J. Tomasi

Volume 39
P. Vanýsek (Ed.)

Electrochemistry on Liquid/ Liquid Interfaces

1985. II, 106 pages. ISBN 3-540-15677-1

Volume 38
E. Lindholm, L. Åsbrink (Eds.)

Molecular Orbitals and their Energies, Studied by the Semiempirical HAM Method

1985. X, 288 pages. ISBN 3-540-15659-3

Volume 37
K. Rasmussen (Ed.)

Potential Energy Functions in Conformational Analysis

1985. XIII, 231 pages. ISBN 3-540-13906-0

Volume 36
I. Ugi, J. Dugundij, R. Kopp, D. Marquarding (Eds.)

Perspectives in Theoretical Stereochemistry

1984. XVII, 247 pages. ISBN 3-540-13391-7

Springer-Verlag
Berlin Heidelberg
New York Tokyo

Volume 35
F. A. Gianturco, G. Stefani (Eds.)

Wavefunctions and Mechanisms from Electron Scattering Processes

1984. IX, 279 pages. ISBN 3-540-13347-1

Volume 34
N. D. Epiotis (Ed.)

Unified Valence Bond Theory of Electronic Structure – Applications

1983. VIII, 585 pages. ISBN 3-540-12000-9

Volume 33
G. A. Martynov, R. R. Salem (Eds.)

Electrical Double Layer at a Metal-dilute Electrolyte Solution Interface

1983. VI, 170 pages. ISBN 3-540-11995-7

Volume 31
H. Hartmann, K.-P. Wanczek (Eds.)

Ion Cyclotron Resonance Spectrometry II

1982. XV, 538 pages. ISBN 3-540-11957-4

Volume 30
R. D. Harcourt (Ed.)

Qualitative Valence-Bond Descriptions of Electron-Rich Molecules: Pauling "3-Electron Bonds" and "Increased-Valence" Theory

1982. X, 260 pages. ISBN 3-540-11555-2

B.-O. Küppers

Molecular Theory of Evolution

Outline of a Physico-Chemical Theory of the Origin of Life

Translated from the German by P. Woolley
1st edition. 1983. Corrected 2nd printing 1985.
76 figures. IX, 321 pages. ISBN 3-540-15528-7

Contents: Introduction. – The Molecular Basis of Biological Information: Definition of Living Systems. Structure and Function of Biological Macromolecules. The Information Problem. – Principles of Molecular Selection and Evolution: A Model System for Molecular Self-Organization. Deterministic Theory of Selection. Stochastic Theory of Selection. – The Transition from the Non-living to the Living: The Information Threshold. Self-Organization in Macromolecular Networks. Information-Integrating Mechanisms. The Origin of the Genetic Code. The Evolution of Hypercycles. – Model and Reality: Systems Under Idealized Boundary Conditions. Evolution in the Test-Tube. Conclusions: The Logic of the Origin of Life. – Mathematical Appendices. – Bibliography. – Index.

The subject of this book is the physico-chemical theory of the origin of life. The theory arises from the application of the principles and methods of physics and chemistry to modern biology. In essence, it provides a physical extension of the classical Darwinian concept of evolution to the molecular level and shows how this new concept can be applied to the problem of the origin of biological information. The monograph is intended as an introductory text also for students of physics, chemistry and biology.

Springer-Verlag
Berlin Heidelberg
New York Tokyo